NATURA
Biology for Bilingual Classes

edited by

Susanna Bächle
Doris Bächle-Knauer

Neurobiology and Behavioural Science

Ernst Klett Verlag
Stuttgart · Leipzig

Warning signs and experiments in school

Natural sciences like biology are not imaginable without experiments. Natura Oberstufe contains also some experiments.

However, experimenting with chemicals is never completely safe. Therefore, it is important to discuss potential sources of danger with your teacher before starting the respective experiment. Especially in the laboratory obvious behavioural rules always have to be observed. The precautions taken are dependent on the danger potential of the used substance.

For that reason the chemicals are marked with a warning sign in the experimental manual. The warning signs are also present on the label of the chemica´s container.

They mean:

C = *corrosive*
Living tissue and materials that come in contact with this substance will be destroyed at the affected site.

F = *flammable*
Substances that are easily ignited when exposed to an ignition source.

Xi = *irritating* (X stands for St. Andrew`s cross)
Substances that can be irritating to skin, eyes or respiratory organs.

Xn = *noxious*
Substances that can cause health problems when inhaled, swallowed or contacted the skin.

---

1. Auflage 1 6 5 4 3 | 19 18 17 16

Alle Drucke dieser Auflage sind unverändert und können im Unterricht nebeneinander verwendet werden.
Die letzte Zahl bezeichnet das Jahr des Druckes.
Das Werk und seine Teile sind urheberrechtlich geschützt. Jede Nutzung in anderen als den gesetzlich zugelassenen Fällen bedarf der vorherigen schriftlichen Einwilligung des Verlages. Hinweis § 52 a UrhG: Weder das Werk noch seine Teile dürfen ohne eine solche Einwilligung eingescannt und in ein Netzwerk eingestellt werden. Dies gilt auch für Intranets von Schulen und sonstigen Bildungseinrichtungen. Fotomechanische oder andere Wiedergabeverfahren nur mit Genehmigung des Verlages.

© Ernst Klett Verlag GmbH, Stuttgart 2010. Alle Rechte vorbehalten. www.klett.de

Bearbeitet von: Susanna Bächle, B. Sc. Molekulare Medizin, Karolinska Institutet, Stockholm; Doris Bächle-Knauer, Max-Planck-Gymnasium, Schorndorf, Staatliches Seminar für Didaktik und Lehrerbildung, Stuttgart

Unter Mitarbeit von: Dr. Irmtraud Beyer, Dreieich; Dr. Horst Bickel, Düsseldorf; Roman Claus, Rees; Roland Frank, Stuttgart; Prof. Dr. Harald Gropengießer, Hannover; Gert Haala, Wesel, Prof. Dr. Siegfried Kluge; Neumark; Bernhard Knauer, Göttingen; Dr. Inge Kronberg, Hohenwestedt; Hans-Peter Krull, Kaarst; Hans-Dieter Lichtner, Stadthagen; Uschi Loth, Burbach; Dr. Horst Schneeweiß, Hamburg; Dr. Jürgen Schweizer, Stuttgart; Ulrich Sommermann, Münchberg; Gerhard Ströhla, Münchberg; Dr. Wolfgang Tischer, Sarstedt; Günther Wichert, Dinslaken

Redaktion: Dr. Peter R. Menke
Mediengestaltung: Ingrid Walter
Fachwissenschaftliche und sprachliche Beraterin:
Dr. R. Theresa Jones, Wolfenbüttel
Lautschrift:
Peter Bereza, Aachen
Layoutkonzeption und Gestaltung
Prof. Jürgen Wirth; Visuelle Kommunikation, Dreieich, unter Mitarbeit von Matthias Balonier, Evelyn Junqueira, Nora Wirth
Umschlaggestaltung
Jens-Peter Becker; normal Industriedesign, Schwäbisch Gmünd; unter Verwendung von folgendem Foto: Jupiter Images GmbH (IFA/Krämer), Ottobrunn/München
Reproduktion
Meyle + Müller, Medien-Management, Pforzheim
Druck
Firmengruppe APPL, aprinta druck, Wemding

Printed in Germany
ISBN 978-3-12-045331-4

What is in this book?

Our knowledge of the natural sciences is constantly developing. Nowadays, the focus has shifted more to the molecular level of processes and many explanations for the "large picture" having been found by the examination of very small molecules and their interactions.

However, the observation and description of the visible world is still a major part of research. The aim of this book is to give an overview of the "large picture" of selected biological principles controlling the way we think (neuroscience) and the way we act (behavioural science).

In order to enrich and aid understanding, this book contains many pages that provide not only information, but also the opportunity to discuss, explore and question scientific topics.

Info pages
These provide basic information about a topic. Numerous figures illustrate complex issues. The short tasks test whether you have understood the text, or not.

CD
The answers to the tasks are on a CD, which also contains word lists (with pronunciation) to help you to learn in a foreign language.

>> **info box** <<

In info boxes you find interesting examples, exceptions, methods, etc.

Extras pages
Extensive information about a specific topic and corresponding tasks are provided on these pages.

Encyclopedia pages
These pages provide additional information which place the topics you have learned about in a broader context and in that way give an overview. Topics are included that are not mentioned directly in the teaching curriculum. These pages are meant to encourage further reading.

Issues pages
Biological problems often go beyond the subject of biology.
These pages provide interdisciplinary material and impulses to work independently. They also refer to topics in everyday life.

Practicals pages
Here experiments are described that you can carry out yourself. It is important to work precisely and to make accurate reports in order to obtain the best results.

Basic concepts
Biology is a complex science and its different disciplines are connected by many, partly abstract, relations. This new type of page shows interconnections between very different topics. Basic concept pages clarify principles and enable you to order and structure known facts. The tasks encourage pupils to discuss and to think about examples again.

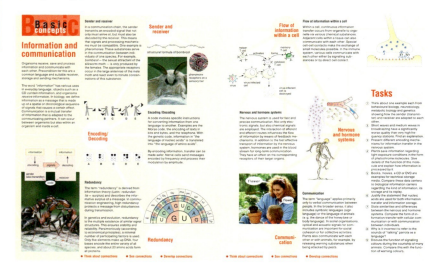

Foreword

GALVANI's discovery of 'animal electricity' in the end of the 18th century caused great euphoria in physicists, physiologists and physicians. However, for thorough research on muscles and nerves, experimental devices were required that were only gradually developed in the 19th century.

In the 1940s the molecular basics of the resting and action potential were revealed. The ion theory of nerve excitation was developed and it was experimentally confirmed that voltage-gated ion channels exist. Today, research is mainly concerned with the complex interactions of millions of individual nerve cells forming a nervous system. How can a nervous system, such as the human brain carry out so sophisticated and complex tasks as seen in behavioural science? Modern imaging techniques enable researchers to observe neuronal processes from the 'outside' and make it possible to connect particular functions to certain areas of the brain. Research with psychoactive substances and their effects on brain function has also become an important approach in solving the secrets of neuroscience.

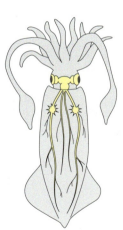

Behavioural science has come a long way from merely 'classical conditioning' to the understanding that behaviour is a product of chemical and physical processes which we are just beginning to discover. The fields of neuroscience and behavioural science are not only very exciting but also essential topics for biology classes in school. Pupils should learn to interpret results of current research and to evaluate them in the light of ethical standards.

After working with this book you should be able to understand the basics principles of neuroscience and behavioural science.

Bilingual classes are challenge for both pupils and teachers. Most scientific articles are written in English, with the language providing a link between researchers around the world and the knowledge that they have. One can say that English has become the "lingua franca" of science. Therefore, it is essential to be comfortable with English and to learn the commonly used scientific terms. This book aims on taking away the scepticism, and maybe even the fear, that teaching and learning in a foreign language might cause. The level of English used in the book corresponds to that expected in the "Oberstufe".

Additionally, pupils should find the word lists for each topic and the accompanying worksheets (on the CD) helpful. The layout of this book corresponds to the chapters 4 and 5 of the German textbook, "NATURA, Biologie für Gymnasien, Oberstufe", which can always be used for reference purposes.

The editors of the book hope that they have succeeded in providing scientific knowledge while helping pupils to learning in English. We wish both the pupils and teachers fun with this book and the fascinating subject of biology.

Susanna Bächle
Doris Bächle-Knauer

April 2010

Contents

Neurobiology

1 Stimulation and conduction 8
The neuron 8
Practicals: The nerve cell 9
The resting potential 10
The action potential 12
Conduction of the action potential 14
From stimulation to action potential 16
Extras: Neurons 17

2 Connections between neurons 18
Synapses 18
Change of coding during information flow 19
Neurotoxins 20
Extras: Medical uses of neurotoxins 21
Neuronal switches — summation processes at synapses 22
Reflexes 24

3 Sensory organs 26
Receptors react to stimuli 26
Encyclopaedia: Human senses 27
The eye as a sensory system 28
Structure of the retina 29
The function of the retina 30
Adaptation: adjustment to light levels 32
Stimulus processing in the retina 33
Receptive fields and contrast 34
Encyclopaedia: Perception 35
From stimulation to sensation 36

4 Structure and function of the nervous system 38
Human nervous system 38
Structure and function of the human brain 40
Encyclopaedia: Methods in brain research 42
Learning: storage — retrieval 44
Issues: Psychoactive substances 46

5 Hormones 48
Transmitter hierarchy 48
Effects of hormones 50
Hormones and development 52
Regulation of the blood sugar level 54
Issues: Diabetes mellitus 56
Stress 58
Plant hormones — phytohormones 60
Hormones and behaviour 62
Extras: Reproductive behaviour of the Barbary Dove 63

Behavioural Science

1 Issues in behavioural science 66
Causal and functional problems 66
The history of behavioural science 68
Encyclopaedia: The instinct doctrine — coming under criticism 69
Methods used in behavioural science 70
Practicals: Wall-seeking behaviour in mice 71

2 Behavioural patterns and their causes 72
Genetically determined elements of behaviour 72
Internal and external motivators 74
Extras: External stimuli driving motion 75
Behavioural sequences 76
Extras: The behaviour of the red-banded sand wasp 77
Reflexes can be influenced 78
Conditioning — an animal as machine? 80
Extras: Models and criticisms thereof 81
Learning and maturation 82
Imprinting 84
Extras: Shrew "caravan" formation 85
Complex learning 86
Encyclopaedia: Other learning processes 87

3 Ecology and Behaviour 88
Habitat choice and territory 88
Optimized nutritional strategies 90
Searching for food 91
Advantages and disadvantages of cohabitation 92
Social systems 93

4 Evolution and Behaviour 94
Successful reproduction 94
Strategies of sexual behaviour 96
Parents invest in their offspring 97
Extras: Reproductive strategies of the dunnock 98
Infanticide and reproductive success 100
Altruistic behaviour 101
Extras: Lifetime strategies 102
Fighting strategies of Red Deer 104
Aggression and social hierarchy 106
Signals and communication 108
Issues: Cultural diversity and universalism 110

Basic concepts
Structures and functions 112
Regulation and control 114
Information and communication 116

Glossary 118
Index 121
Pictures sources 126

Neurobiology

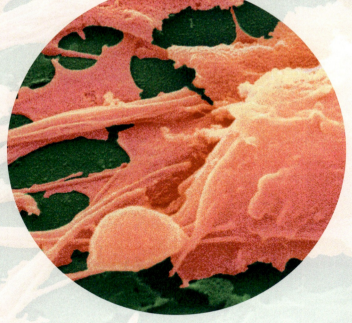

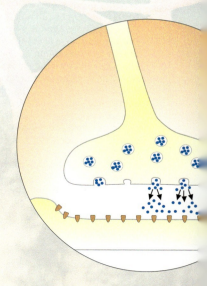

The stimuli initiate electric impulses in the sensory organs. Sensations such as the smell and colour of the salmon arise from the excitation of nerve cells in specific areas of the brain.

A bear is hunting for salmon. The smell of the fish has attracted the bear and now it is standing in shallow water, waiting and watching. The bear's paw strikes just at the right moment. The bear is successful because it is experienced in salmon hunting. The processes in the bear's nervous system during salmon hunting are presented here in a simplified way. A more precise examination of such processes is the task of *neurobiology*.

The bear can only exist if it extracts information from its environment, processes it and reacts to it. Stimuli from the environment stimulate specific sensory cells in the sensory organs.

The brain processes the numerous stimuli, saves them and compares them with previously saved information.

Sensory cells are selective and react only to appropriate stimuli. The sensory cells in the nose react to volatile chemical substances, whereas the eye reacts to light. Such stimuli generate signals, which are transported to the brain by thread-like nerves. These nerve fibres are called *sensory* or <u>afferent</u> fibres because they forward impulses to the brain.

The brain and spinal cord are together referred to as the *central nervous system* (CNS). The CNS not only receives stimuli, but also processes them and, in our example, influences the bear's behaviour when catching the fish. The observable behaviour is the sum of many coordinated muscular movements. In order to do this, the CNS sends signals via nerves to the muscles of the bear's paw. These nerves are called <u>efferent</u> or *motor* nerves because they forward impulses from the CNS to the muscles.

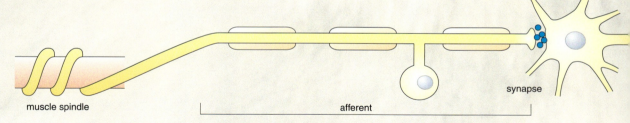

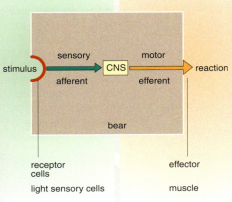

Not only muscles, but also inner organs and glands, for example the salivary glands, are stimulated by motor nerves. Structures that are influenced by efferent nerves and show a reaction are called *effectors*.

Changes in brain activity can be measured and assigned to specific brain regions. Changes of state within the body, such as an alteration in blood glucose level (a measure of hunger), and stored information are compared by the brain. A hungry bear will start hunting in a region where it has previously been successful.

efferent

neuromuscular junction   muscle

effector

1 Stimulation and conduction

The neuron

Nerve
Bundle of nerve fibres surrounded by connective tissue.

Nerve fibre
Axon with surrounding sheath cells.

Neuron
Mostly branched cell with long extensions specialized in the processing and conduction of impulses.

The most important parts of a nervous system are the nerve cells also called _neurons_. These cells can produce, process and transmit electrical impulses. The specialisation of the various neuron types can be seen in their shape, branching and degree of extension (see margin). Their length can vary from a few micrometers to more than one meter. However, neuron diversity can be traced back to a uniform and typical building plan (fig. 1).

The neuron is divided into a _cell body_ (soma) and cellular extensions. The _soma_ contains, amongst other things, the nucleus. The cell extensions of each neuron can be classified as _dendrites_ and an _axon_. Dendrites often form extensions with many branches ("dendritic tree"), which are usually not longer than 2 mm. Close to the soma, they are thicker than the axon but become thinner following each fork. The axon is often much longer than the dendrites. A neuron has only one axon with a cone-shaped bump (_axon hillock_) from which it originates. Dendrites transmit impulses towards the soma and axons send impulses away from it. Many axons are branched distally. Each axon terminal exhibits a swelling. These _synaptic knobs_ represent functional connections (_synapses_) either to muscle fibres or to other neurons via their dendrites or soma.

Neurons are surrounded by sheath cells called _glial cells_. It is estimated that there are about 10 times more glial cells than neurons. They provide support, electric insulation and nutrition for the neurons. The axons of the sensory and motor nerve cells are surrounded by cell membrane layers of specialised sheath cells (_myelin_) in many vertebrates. They form the so-called _myelin sheath_ by entwining around the axon during its development (fig. 2). An axon and its surrounding sheath cells are called a _nerve fibre_. Bundles of many of these fibres are enclosed by connective tissue and form a thick _nerve_. It looks white through a light microscope because of the many lipids in the cell membranes of the sheath cells. The somas (also: somata) of nerve cells usually have a grey appearance.

Neuron in the cerebellum

in the spinal cord

in the autonomic nervous system

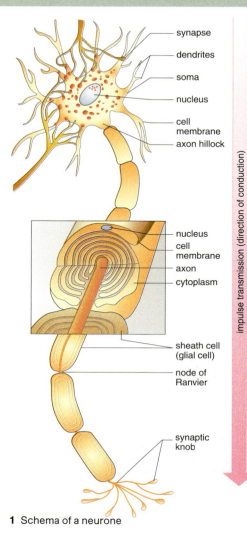
1 Schema of a neurone

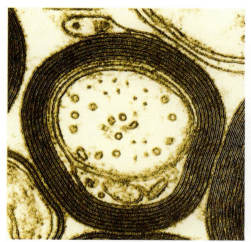
2 Axon with myelin sheath

8 Neurobiology

The nerve cell

Preparation of nerve cells

Material:
— Spinal cord of a slaughtered animal (e. g. pig), fresh, slightly frozen on the surface
— sharp razor blade or scalpel
— scissors or tweezers with sharp points
— 6 microscope slides
— 2 Petri dishes
— microscope and equipment
— Giemsa solution, distilled water

Experiment
a) Cut the spinal cord crosswise and re-form its round structure. A connective tissue sheath surrounds an outer white and an inner slightly pink butterfly-shaped area. These areas represent the white and grey matter, respectively.

b) Remove some grey substance from the anterior horn by using the scissors or the tweezers and place it onto a microscope slide.
c) Repeat experiments a) and b) until 5 samples have been placed on microscope slides.
d) The material must now be squeezed by using another microscope slide and expanded by moving the two slides in opposite directions (see diagram below).
e) Add a thick layer of *Giemsa solution* to stain the sample and incubate the preparation for 5 min (in a Petri dish).
f) Wash the excess Giemsa solution away carefully with distilled water.

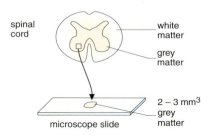
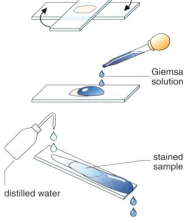

g) Examine the samples with the microscope: the cell body and cell extensions are stained purple. Note the nuclei with their dark-contrasted nucleoli and the irregularly shaped endoplasmic reticulum. The nuclei of the sheath cells can also be seen.

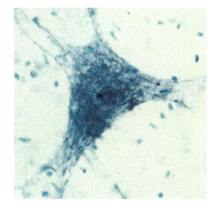

In 1885, CAMILLO GOLGI discovered that only a few cells are stained when treating nerve cells with specific heavy metal salts. Some cells stand out in the dense nervous tissue of the brain. The spatial arrangement of the dendrites and axons can be detected by this method.

SANTIAGO RAMÓN Y CAJAL (1852 – 1934) used the Golgi staining method to examine the nervous systems of humans and many other vertebrates. Even today, his drawings form the basis of neurobiology.

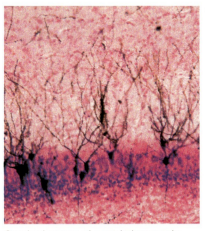

Cerebral cortex of a cat (micrograph; Golgi stain)

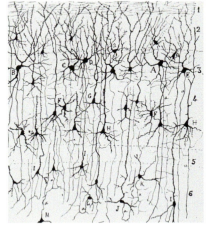

Cerebral cortex of a rat (line-drawing of CAJAL 1888; Golgi stain)

Task

① Compare the two figures (micro and drawing) with regard to their information value. Why do the axons not appear in their full length in the micrograph?

Neurobiology

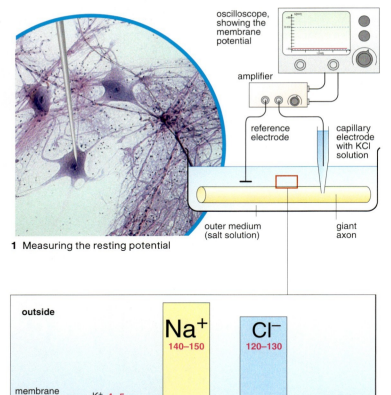

1 Measuring the resting potential

2 Axon with ion distribution (ion concentration in mmol/l)

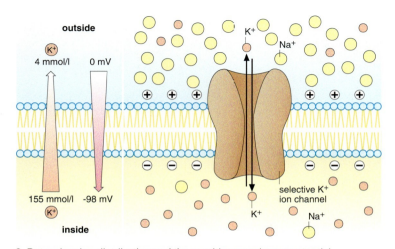

3 Potassium ion distribution and the resulting membrane potential

The resting potential

As early as the 18th century, LUIGI GALVANI observed contraction in the nerve-muscle tissue of a frog leg while he was preparing it with metal instruments. He explained this observation on the basis of "animal electricity". More precise explanations could only be proposed after 1936 following the development of equipment that made it possible to detect and quantify (i.e. measure) extremely weak voltages and currents.

Membrane potential

In order to study the electrical processes in a neuron, single axons rather than the entire nerve are examined. Measurements are performed by using electrodes that are connected, through an amplifier, to an oscilloscope. To carry out measurements within the thin axons, extremely fine hard glass capillaries have been developed that are filled with a conducting salt solution.

If a non-stimulated axon is measured by locating both electrodes outside the membrane in the extracellular body fluid (*extracellular recording*), no voltage is detected. If one electrode is inserted through the axon membrane, a voltage is detected between the two electrodes as shown on the oscilloscope (fig. 1).

Voltage is the difference between two electrical potentials. When examining an axon, a voltage is measurable between the outside and the inside of the axon membrane. This voltage is called the *membrane potential*. The potential of the outer reference electrode is defined arbitrarily as being zero and, hence, a potential of $-70\,mV$ exists inside an axon. Membrane potentials exist not only in neurons, but also in all plant and animal cells (e.g. muscle cells).

A voltage develops across the cell membrane because positively and negatively charged ions are present at both sides of the membrane, but in different concentrations (fig. 2). These ions can only pass through the membrane in a controlled manner. The *resting potential* is based on the distribution of positively charged K^+ and Na^+ ions and negatively charged Cl^- and organic (A^-) ions. The organic anions are acidic residues of organic acids or proteins.

Generation of the resting potential

The axon membrane consists of a double layer of lipids, as is the case for all membranes. It separates the cytoplasm in the intracellular space from the body fluids (*lymph*) in the extracellular space. The membrane is selectively permeable. Water molecules can pass through it unrestrictedly, dissolved ions cannot pass through it or can only move in a restricted manner.

Experiments in which radioactively marked ions were added either to the intracellular or the extracellular space, however, showed that some of these ions are found later on the alternative side of the axon membrane. The permeability for the particular ions is different (see margin): for Na^+ ions, it is only about 4% of the permeability of K^+ ions. Charged protein molecules cannot pass the membrane at all. This selective permeability of the axon membrane can be explained on the basis of the presence of *selective ion channels*. Various channel proteins have been discovered that are located within the lipid layer of the cell membrane. They only let one type of ion pass, e.g. K^+ ions (fig. 10.3).

In a resting state, mainly the K^+ ion channels are opened in the axon membrane. The concentration of K^+ ions inside the axon membrane is 40 times higher than outside the axon. The probability that a K^+ ion passes a channel from the inside to the outside is 40 times higher than the other way round. Each K^+ ion that moves through the ion channel from the inside to the outside removes a positive charge from the axon interior. The intracellular space is thus charged negatively relative to the extracellular space, because the number of positively charged ions decreases. The negatively charged ions remain. Since oppositely charged ions attract each other, negatively charged ions accumulate close to the inside and positively charged ions close to the outside of the axon membrane. This charge separation causes the development of an electric field across the membrane.

Because of the excess positive charge on the outside, the following K^+ ions are repelled increasingly by the identical charge. Therefore, two forces influence the K^+ ions (fig. 10.3):
— The *concentration gradient* between inside and outside that promotes the exit of K^+ ions through the ion channels.

Ion permeability
(relative values)
K^+ 1
Na^+ 0.04
Cl^- 0.45
A^- 0

Ion diameter including water shell:
K^+ 396 pm
Na^+ 512 pm

1 pm = 10^{-12} m

Resting potential (mV)
cat   −60 to −80
(motoneuron)

crab   −71 to −94
(giant axon)

squid   −62
(fibre without myelin sheath)

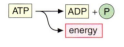

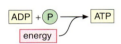

ATP
(adenosine triphosphate)
a high-energy substance

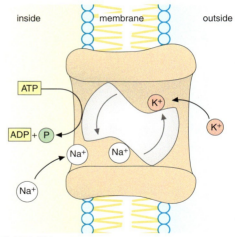

1 Na^+/K^+ pump

— The electric field caused by the charge separation that promotes the opposite entry of K^+ ions.

This finally leads to a state in which both processes are in equilibrium. The voltage hereby developed by the K^+ ions, i.e. the potassium equilibrium potential, is the basis for the membrane potential. It is present in a non-stimulated or resting neuron. This voltage is therefore called the resting potential. The strength of the resting potential is influenced only a little by movements of Na^+ and Cl^- ions.

Sodium-potassium pump

The resting potential should remain in equilibrium for a long time. However, if the production of the high-energy substance ATP is hindered by cell toxins, then the resting potential slowly degrades. Small amounts of Na^+ ions diffuse to the inside and subsequently K^+ ions diffuse to the outside because of the concentration gradient and the membrane potential. These leakage currents through the axon membrane are balanced by an active energy-consuming transport mechanism called the *sodium-potassium pump* (Na^+/K^+ pump, fig. 1). Without it, the ion concentrations would slowly even out and the resting potential would reach zero. The pump is a membrane protein that transports Na^+ ions out of the cell and K^+ ions into the cell. The energy required for this process is delivered by ATP. In one cycle, the ion pump exchanges three Na^+ ions for two K^+ ions (*electrogenic pump*).

Neurobiology

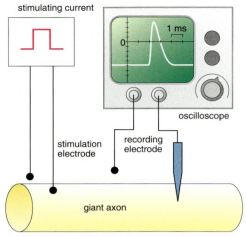

1 Measurement of an action potential

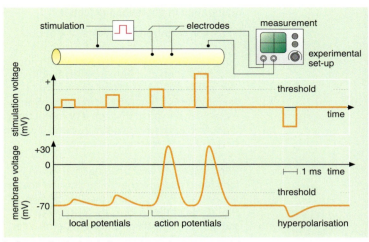

2 Strength of stimulation and membrane potential

The action potential

Stimuli cause electric impulses in sensory cells. Sensory nerve cells transmit these signals to the brain. Motor nerve cells conduct impulses to the muscles where they cause contractions. A short-term change of the membrane potential can be measured on the axon membrane of the responsible nerve cells at the moment of transmission.

Measurement of the action potential

The above relationships have been examined scientifically since 1936 by using the axons of *Loligo*, a squid. The axons are especially large in diameter (up to 1 mm) and therefore are highly suitable for measurements.

In experiments, the axon is stimulated by the application of various voltages at a specific site. On a site located a small distance away from the application point, the reaction of the axon is measured with a glass capillary and an oscilloscope.

If a voltage is applied by the stimulating electrodes for a short period of time and if this voltage is more negative than the resting potential, then the voltage also decreases at the recording site (*hyperpolarisation*). Stimulation with a voltage of the opposite polarity results in a short increase in the axon potential (*depolarisation*). The higher the stimulating voltage, the higher is the depolarisation at the recording site. If the depolarisation exceeds a specific

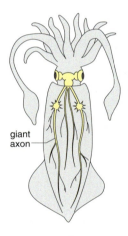

giant axon

Hyperpolarisation
hyper, Greek = excessive; voltage is more negative than the resting potential

Depolarisation
de, Latin prefix = away from; fast change in the resting potential to a more positive voltage

Repolarisation
re, Latin prefix = back; return to the resting potential

threshold voltage, the axon potential changes abruptly within one millisecond reaching a top value of +30 mV. The axon actively produces an *action potential* that travels along the axon. A fast and strong increase in the potential is measured. This is called the *rising phase* or *depolarisation phase*. During the *overshoot*, the axon potential can reach positive values. After a short time, the potential sinks once again to the level of the resting potential during *falling phase* or *repolarisation*.

Action potentials always show the same course. The duration of the single phases and their electric potentials are always the same. They happen fully or not at all and thus strictly follow the *all-or-nothing principle*.

Molecular processes

A change in the membrane potential can be explained by a change in the ion concentrations on the inside and outside of the axon membrane. In 1949, ALAN HODGKIN and BERNARD KATZ carried out experiments on the axons of squids. They replaced the Na^+ ions present on the outside of the axon with choline ions, which are also positively charged but are significantly larger and thus are not able to pass through the axon membrane. Under these conditions, no action potential could be obtained. The researchers deduced that the action potential was produced by the opening of Na^+ ion channels and the fast entry of

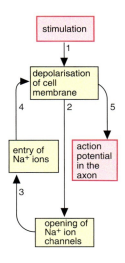

Na$^+$ ions into the axon (*ion theory of excitation*). The measurement of the ion concentrations present showed that the outer concentration of Na$^+$ ions was 10 times higher than the inner concentration. The hypothesis was confirmed by the direct measurement of ion currents at the ion channels: during depolarisation, most Na$^+$ ion channels are open but only a few are opened for K$^+$.

Figure 1 demonstrates the development of an action potential. The axon is depolarised past its threshold by an excitatory stimulus. Hereby, nearby Na$^+$ ion channels are opened (fig. 1b). These channels are *voltage-gated channels*. A great number of Na$^+$ ions enters the inside of the axon. The fraction of positively charged ions thus decreases outside the axon but increases inside. This change in the membrane potential leads to the opening of further voltage-gated Na$^+$ ion channels and so to further depolarisation until the maximum for the action potential is reached (fig. 1c). The individual Na$^+$ ion channels close fast and cannot be opened again for a short period of time. They only become active after returning to the resting potential following a *refractory period*.

The repolarisation of the axon membrane is mainly based on the release of K$^+$ ions (fig. 1d). This can be shown in experiments with toxins that block voltage-gated K$^+$ ion channels. Repolarisation is extremely slow in these experiments. Measurements of voltage-gated K$^+$ ion channels have shown that their opening is delayed after depolarisation. The increasing release of K$^+$ ions out of the axon alters the charge inside of the cell to a more negative value until the resting potential is reached once again. The K$^+$ ion channels close only once the resting potential has been regained.

A single action potential changes the ion concentrations at the membrane only slightly so that the resting potential is easily reached again. If thousands of action potentials travel along the axon membrane, the importance of the sodium-potassium pump rises. It prevents the ion distribution changing to such an extent that no resting potential can occur again.

Task

① Explain why the depolarisation that leads to an action potential is referred to as self-perpetuating process (see margin).

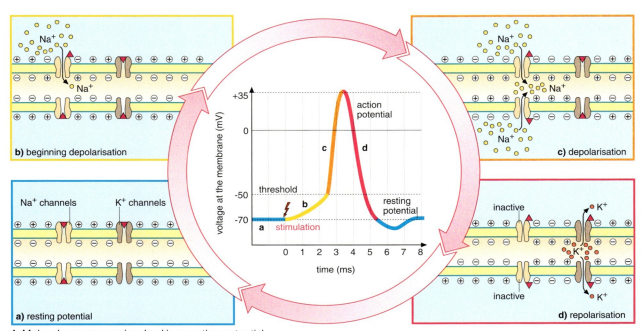

1 Molecular processes involved in an action potential

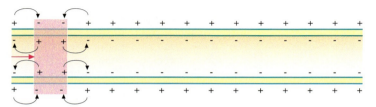

1 Continuous conduction

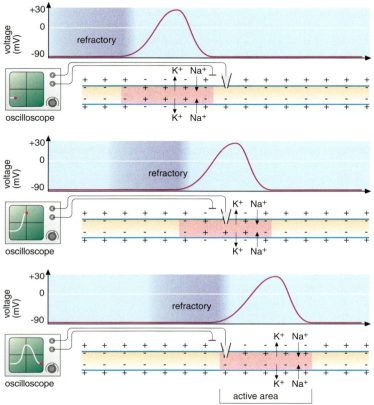

2 Snapshots of the conduction of an action potential

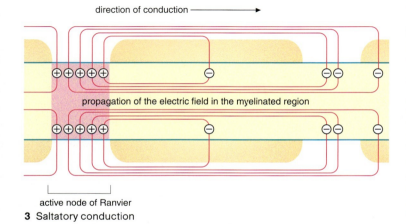

3 Saltatory conduction

Conduction of the action potential

An electric stimulation below the threshold causes a depolarisation of the axon at the site that was stimulated. The stimulated site has a more positive charge at the inner membrane than in its environment. An electric field is produced in both directions along the axon (fig. 1). The field strength, and thus the depolarisation of the membrane, decreases with increasing distance. High electrical conductivity along the axon membrane leads to further expansion of the electric field. The same effect has a preferably small conductivity across the membrane so that the electric field cannot be weakened by ion flow through the membrane. The electric field can expand (also: propagate) between 0.1 and 1 mm depending on the conditions. Such a passive expansion (propagation) of the voltage change along an axon is called *electrotonic conduction*.

Continuous conduction

In contrast to local potentials, action potentials are conducted along the axons without getting weaker. If the voltage threshold is exceeded across the axon membrane, an electric field is produced that is strong enough to exceed the threshold in other regions of the axon and the voltage-gated sodium ion channels are opened. The resulting depolarisation triggers another action potential that again causes a depolarisation above threshold at the neighbouring site. The continuous repetition of these processes conducts the action potential further and further along the axon (fig. 2). Since it is newly produced according to the all-or-nothing principle at the various membrane sites, it is not weakened with increasing distance. This is called *continuous conduction*.

The electric field can theoretically propagate in both directions along the axon. Nevertheless, the axon potential usually only propagates in one direction. At the site where an action potential has just been produced, the axon membrane is in the *refractory period*. At this time, the sodium channels are inactive and cannot be opened. Only the sodium channels located in front of the action potential can be activated and thereby opened to trigger an action potential. Thus, in an organism, an action potential always runs from the axon hillock to the synaptic knobs. Axons are "one-way streets".

14 Neurobiology

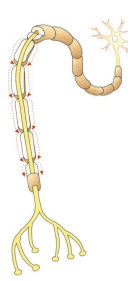

Saltatory conduction

Saltatory
lat. *saltare* = to hop or leap

Fast reactions, for example during hunting or while escaping, ensure the survival of an animal. During evolution, two mechanisms for the rapid conduction of action potentials have been developed. The increase of axon diameter lowers the electric resistance along the axon membrane. During depolarisation of the neighbouring sites by an electric field, the threshold is reached faster. Hereby action potentials are triggered faster and the conduction speed is higher than in thinner axons.

Saltatory conduction

An extremely high conduction speed is found in *myelinated axons*. Action potentials are only produced at the non-myelinated *nodes of Ranvier* because voltage-gated sodium ion channels are only found here. The myelinated regions seem to be "jumped over" by the action potentials, almost without any loss in time. This is therefore referred to as <u>saltatory conduction</u> (fig.14.3).

The membranes of the glial cells are wrapped around the axon forming many layers. The lipids of the membranes (myelin sheath) are good electrical insulators and provide an extremely low conductivity through the membrane. Thus, the electric field expands quickly along the axon during depolarisation and is weakened only slightly. It therefore depolarises the membrane at the neighbouring node very rapidly.

Because of the high density of sodium channels at the node (up to 12,000 per μm^2 instead of "only" 500 in a non-myelinated neuron), the change of voltage during the production of the action potential occurs at high speed. In human axons, a conduction speed of up to 180 m/s can be observed. This is because the nodes of Ranvier lie about 2 mm apart and the time-consuming amplification to the threshold level only has to take place at these nodes.

Saltatory conduction is not only superior to continuous conduction with regard to time. Since a smaller axon diameter is sufficient to allow such conduction, material is saved. Only thin axons enable the development of complex nervous systems in a restricted space such as the brain. Additionally, the energy consumption is also low.

Tasks

1. Explain why saltatory conduction requires less energy than continuous conduction.
2. In an experiment, the middle of a prepared axon is stimulated above threshold. How does conduction differ in this situation compared with that under natural conditions?

»info box«

Search for ion channels — the patch-clamp technique

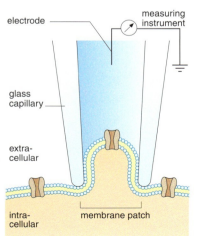

In 1991, the two German scientists E. Neher and B. Sakmann received the Nobel Prize for the development of this technique. It can be used for all kinds of cells, although in plant cells, the cell wall has to be removed first. The cell is attached by suction to a holding pipette. A tiny area of the membrane (the so-called "patch") is isolated by means of a micropipette and the ion movement through a single ion channel or an ion pump can now be analysed. An artificial current is used to balance the natural ion currents and stabilizes the cell potential at the desired level (the so-called "clamp"). The artificial current is measured; this is reversely equal to the natural current through the membrane. Hereby, during an action potential, a sodium channel was found to be open for about 0.7 ms and to allow the passage of about 10,000 ions.

From stimulation to action potential

Stimuli from the environment and the body must be translated into the language of the nervous system. In animals and humans, specialized neurons function as *receptor cells*; these are called *primary sensory cells*. An example are the *muscle spindles* that lie parallel to the muscle fibres in our muscles (fig. 1). These react to contractions of single muscle fibres and regulate muscle contraction during movement.

If the muscle fibre is stretched, the muscle spindles are also stretched. This causes an opening of Na+ ion channels in the sensory neuron. Na+ ions can diffuse into the axon and the membrane is depolarized. This depolarisation is called the *receptor potential*. In this specialized region, no voltage-gated Na+ ion channels are present that can trigger an action potential.

Voltage-gated ion channels are only found in the axon. The receptor potential is therefore a localised potential and is further conducted electrically. The electric field propagates over the membrane. An action potential is triggered, and thereafter conducted, only once the threshold is reached in the region where voltage-gated Na+ ion channels are located. Figure 1 demonstrates it.

High-intensity stimuli, e. g. extensive stretching of the muscle fibre, cause a higher amplitude of the receptor potential than a low-intensity stimulus. This is because high-intensity stimuli result in the opening of more ion channels through which Na+ ions can diffuse. During the initial phase, the amplitude of the receptor potential increases linearly with the intensity of the stimulus (see margin). The amplitude and duration of the receptor potential determine the frequency and period of the initiated action potentials. In the simplest case, a stronger stimulus leads to a higher frequency of action potentials. Therefore, the stimulus intensity is encoded in frequency (fig. 1). When the stimulus lasts for a long period of time, the amplitude of the receptor potential decreases and consequently the action potential frequency also decreases.

Tasks

① Why is it not possible to apply the all-or-nothing principle to the receptor potential?
② Look at the figures and explain the connection between stimulus intensity, receptor potential and action potentials and describe the influence of stimulus duration.

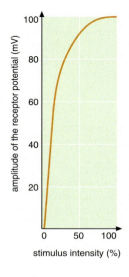

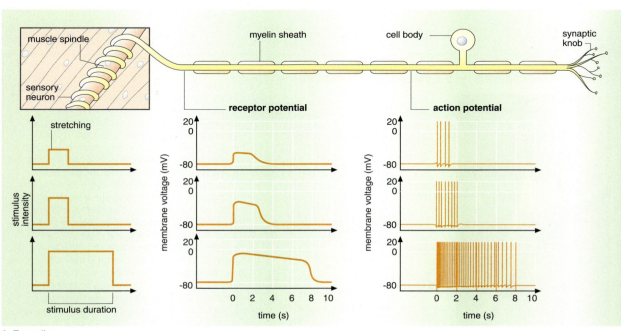

1 Encoding

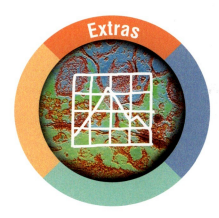

Neurons

K⁺ concentration and resting potential

In an experiment, the extracellular K^+ concentration can easily be changed by bathing the axon in different solutions. The resting potential can be measured for each solution.

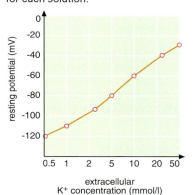

1 Influence of K^+ concentration

Task

① Describe the results (fig. 1) and explain their molecular background.

Type of nerve fibre	Fibre diameter (μm)	Median conduction speed (m/s)	Examples
non-myelinated	1	1	slow nociceptor (mammal)
	700	25	giant fibre (squid)
myelinated	3	15	sensory fibre of mechanoreceptors of the muscle (mammal)
	9	60	sensitivity to touch of the skin
	13	80	sensory fibre of muscle spindles (mammal)
	13	30	fibre in the spinal cord (frog)

Influence of Na⁺ concentration

The Na^+ concentration in the solution bathing an isolated axon is gradually decreased. For this, the Na^+ ions are replaced by glucose molecules. Action potentials are triggered and recorded. In figure 2, the changes are shown compared with the amplitude of an action potential under normal conditions.

Tasks

② Draw a typical action potential and mark the changes that are caused by the decrease of the extracellular Na^+ ion concentration.
③ Explain the observations by using the ion theory of the action potential.
④ What would happen if the Na^+ ions were not replaced by glucose molecules?

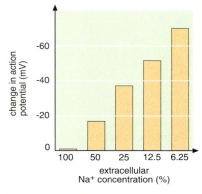

2 Influence of Na^+ concentration

Poisoning neurons

Cyanides are highly toxic because, amongst other things, they block the respiratory chain so that no ATP is available. If cyanides are added to neurons, action potentials are initially triggered. After a while, however, action potentials can no longer be triggered. Additionally, the resting potential is decreased.

Task

⑤ Explain the effect of cyanides on neurons.

Conduction speed

The conduction speed of a neuron in the forearm is measured by applying a small electric shock to the elbow and after that to the wrist (fig. 3). The sites are about 27 cm apart from each other. The effect of the stimulated action potentials is measured extracellularly as muscle action potential in the thumb muscle. Muscle cells also show an action potential.

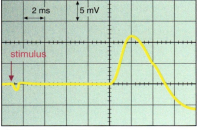

a) after stimulus at the elbow

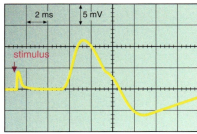

b) after stimulus at the wrist

3 Conduction speed

Tasks

⑥ Calculate the conduction speed from the given values.
⑦ In the table on the left, examples for the average conduction speed of different neurons are presented. Determine from these data the factors that influence the speed of conduction and explain them.

2 Connections between neurons

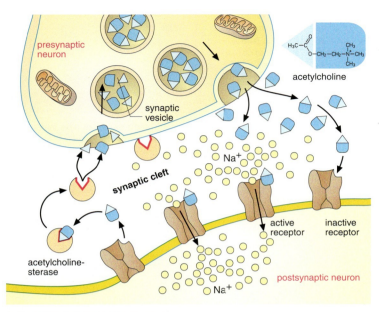

1 Functional diagram of a synapse

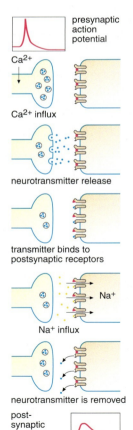

Many receptor molecules are located in the postsynaptic membrane; these fit the neurotransmitter molecules like a lock and a key. Receptor and neurotransmitter bind for a short period of time leading to a change in the conformation of the receptor protein. In the case of acetylcholine, this causes the opening of a Na^+ ion channel that is linked to the receptor molecule. Sodium ions enter the postsynaptic cell and depolarise it. The more neurotransmitter molecules are released, the more the ion channels are opened and the greater is the resulting depolarisation.

After a short time, the transmitter molecules are cleaved by an enzyme and the inactive cleavage products are taken up into the synaptic terminal. They are used for the synthesis of new neurotransmitter. If the neurotransmitter was not removed from the synaptic cleft, it could cause permanent excitation of the postsynaptic nerve cell.

Few voltage-gated ion channels are present in the region of the cell body. They mostly occur only in the axon and the axon hillock. If the depolarisation in the postsynaptic cell reaches a certain level, it can trigger an action potential, which is therefore called an **e**xcitatory **p**ost**s**ynaptic **p**otential (*EPSP*).

Synapses

The site connecting one neuron with another one or a neuron with a gland or muscle cell is called *synapse*. Here, impulses can be transmitted. A synapse is made up of an enlarged axon terminal, the <u>synaptic knob</u>, a cleft called the <u>synaptic cleft</u> (only visible under an electron microscope) and the opposing region of the membrane on the neighbouring cell. The cell and those parts in front of the synaptic cleft are referred to as being *presynaptic*, whereas those on the other side of the cleft are termed as being *postsynaptic*.

Membrane-enclosed vesicles called *synaptic vesicles* are found in the synaptic terminal. They contain tiny amounts of a transmitting substance (*neurotransmitter*), e.g. *acetylcholine*. If the action potential reaches the synaptic terminal of a neuron, voltage-gated calcium ion channels open and calcium ions can enter the cell along a concentration gradient. This causes the synaptic vesicles to migrate to the presynaptic membrane with which they fuse. Neurotransmitter molecules are released into the synaptic cleft and are distributed there by diffusion. Therefore, the transmission of an action potential is not electrical but chemical.

Tasks

① Explain the difference between sodium ion channels in the axon membrane and in the synaptic cleft.

② Why are the neurotransmitter molecules cleaved so fast in the synaptic cleft (50 molecules/ms)?

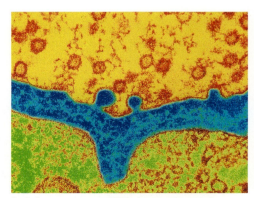

2 Colour-contrasted EM picture of a synapse

Neurobiology

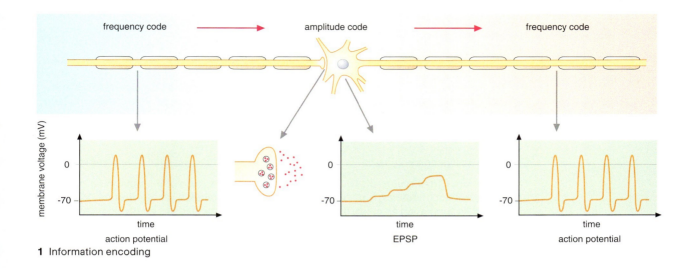

1 Information encoding

Change of coding during information flow

Task

① Relate the terms analogue and digital coding to the diagram.

In the axon, information is transmitted as action potentials. The duration and intensity of the stimulus are encoded as the number of action potentials per time (frequency). This is therefore a *frequency code*.

If an action potential reaches the synapse, neurotransmitter molecules are released into the synaptic cleft and the postsynaptic cell becomes depolarised. The faster the action potentials arrive after each other, the more transmitter molecules are released and the higher is the resulting EPSP. The information encoded in the frequency of action potentials is now newly encoded into transmitter release and the EPSP.

As here the amplitude representing the amount of neurotransmitter and the height of the potential is variable, it is now called an *amplitude code*. In the axon hillock, the EPSP once again triggers potentials that are then transmitted. During the conduction of information from one neuron to the next, the information changes its code several times.

Information conduction encoded in frequency is more secure than the amplitude code because the height of the potential can easily be weakened if transmitted over long distances. This would distort the information.

»info box«

Neuromuscular junction — motor end plate

Action potentials are transmitted via a motor neuron to the muscle. The muscle then contracts. The link between the nerve cell and the muscle is a specific synapse called the *motor end plate*.

Motor end plates are larger than synapses between two neurons. However, they have the same basic structure. Here, neurotransmitter molecules are released (in this case, acetylcholine) that cause the opening of sodium channels on the postsynaptic membrane (in this case, the muscle fibre membrane). The sodium influx leads to the depolarisation of the muscle fibre membrane. This initiates an action potential that propagates along the muscle fibre membrane.

Action potentials provoke the influx of calcium ions into the plasma of the muscle fibre cell. The free calcium ions enable the binding of *actin* and *myosin*. The controlled sliding of these filaments along each other leads to the contraction of the muscle fibre.

Neurobiology **19**

Neurotoxins

Plants and animals often find it advantageous to produce toxins that cannot only be used against herbivores and predators, but also for obtaining prey. Much is known today about the fast-acting toxins that affect the motor end plate and thus the conduction between nerve and muscle cell.

Change in acetylcholine release

The bacterium *Clostridium botulinum* lives under anaerobic conditions, e.g. in badly conserved tin food such as meat, fish or beans. Contaminated tins can be recognized by their curved lid, which has been pushed up by the gases produced during fermentation. *Botulinum toxin* is one of the strongest toxins known: 0.01 mg in food or 0.003 mg in the blood circulation is deadly. Symptoms such as headache or "muscle fatigue" occur after 4 to 24 hours. Death is caused by respiratory paralysis or cardiac arrest. The toxin is made ineffective when boiled. Botulinum toxin degrades a protein that is present in the membrane of the synaptic vesicles; this protein is important for fusion of the synaptic vesicle membrane with the presynaptic membrane and is thus required for the release of the neurotransmitter acetylcholine into the synaptic cleft. The release of acetylcholine is hence inhibited. Action potentials can no longer be transmitted from the nerve to the muscles. Botulinum toxin is nowadays used in medicine against abnormal muscle cramps and even in cosmetic medicine to remove or prevent the formation of wrinkles in the skin.

The toxin of the *black widow*, a spider belonging to the genus *Latrodectus*, causes chills, pain and shortness of breath. Sometimes, death occurs because of respiratory paralysis. The toxin provokes the simultaneous release of all synaptic vesicles of the motor end plates into the synaptic cleft.

Blockage of the acetylcholine receptor

Coniine, the toxin of the poison *hemlock* (*Conium maculatum*) causes limp (floppy) paralysis and leads to death from respiratory paralysis while the victim is fully conscious. PLATO described the death of SOCRATES who was forced to drink a *cup of hemlock*. The active component binds reversibly to the receptor molecules for acetylcholine without opening the sodium ion channels.

Suxamethonium, a substance similar to acetylcholine, causes cramping because of permanent depolarisation. It opens the sodium ion channels but is degraded significantly more slowly by acetylcholinesterase than is acetylcholine.

Inhibition of acetylcholinesterase

Alkyl phosphates are organic esters of phosphoric acid and a component of insecticides and plasticizers in plastics and chemical warfare agents (*tabun*, *sarin*). They inhibit acetylcholinesterase irreversibly. This causes cramping of the skeletal muscles because of permanent depolarisation and leads to death by respiratory paralysis.

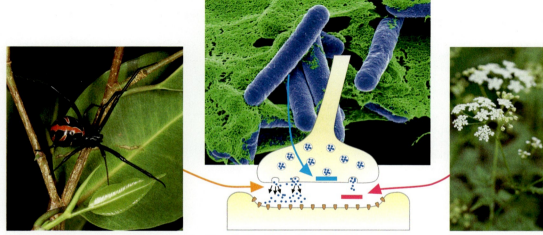

1 *Clostridium botulinum* (middle), black widow (left) and hemlock (right)

Neurobiology

Medical uses of neurotoxins

Myasthenia gravis

Myasthenia gravis means severe muscle weakness. The patients suffer from the fact that their skeletal muscles fatigue easily. This gets worse during the course of the day and especially under physical exercise. However, it gets better after resting.

The eyelids of the patients characteristically drop and their facial expression is disrupted. Some patients have problems in talking. In severe forms, the symptoms also affect shoulders, arms and legs. Unusual myasthenic crises can lead to death caused by the failure of muscles involved in swallowing and breathing.

About 5 to 7 out of 100,000 people are affected. In these patients, the conduction between nerve and muscle cells is disturbed. The cause is a so-called *autoimmune disorder*: The immune system produces antibodies against acetylcholine receptors so that these are blocked. Based on this knowledge, new therapies are being developed. The therapeutic options work well and many patients are free of symptoms and can work as normal.

Tasks

① Explain the connection between the production of antibodies directed against the acetylcholine receptor and the occurrence of the symptoms described above.
② Based on this, suggest potential medication.
③ Gather information, e.g. from the Internet, about current therapies for myasthenia gravis and explain why they are effective.

Curare — site of action

Curare is a mixture of various plant toxins that is used by natives in South America on the tips of their spears. If the toxin passes into the blood stream of prey, the skeletal muscles are paralysed. Poisoning is prevented by heating the meat before eating. The toxin degrades during the heating process.

In order to solve the question regarding the site of action of curare, the following historical experiment was conducted in 1857 by CLAUDE BERNARD.

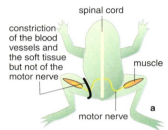

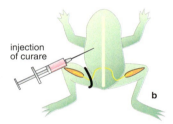

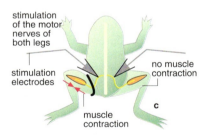

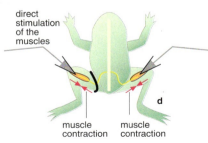

Tasks

④ Describe the experiment and its results.
⑤ What conclusion can you draw about the site of action of curare? Explain.

"Neurotoxins" in medicine

The isolated active agent of *tubocurarine* is used in operations to paralyse the muscle by totally relaxing it. This makes it possible, for example, to stop the breathing movements of a patient while the surgeon operates on the chest.

When the operation is finished, the muscle relaxing effect can be reversed by injecting a substance called *neostigmine*. Neostigmine is an alkyl phosphate that binds reversibly to acetylcholinesterase and inhibits its function only for a short period of time.

Tasks

⑥ What properties of a synaptic inhibitor are essential for therapeutic use?
⑦ What reasons can be found not to use alkyl phosphates to reverse the effect of tubocurarine?
⑧ What can you assume about the exact site of action of curare and tubocurarine based on the knowledge regarding the effect of neostigmine? Explain.

Atropine is the toxin of the *deadly nightshade* (*Atropa belladonna*) and other plants within the family *Solanaceae* (the nightshade family). It binds not only to the sodium ion channels in the synapses of the heart and other internal organs, but also to the muscles of the iris in the eye. When the iris constricts, the pupil becomes narrower.

Tasks

⑨ Why are drops of atropine often placed in the eye before an eye examination?
⑩ Dilated pupils indicate attention to other people and make a person seem more "likeable". Relate the term "belladonna" to the toxin of the deadly nightshade.
⑪ Atropine is used as antidote in cases of poisoning with an acetylcholinesterase inhibitor. Explain.

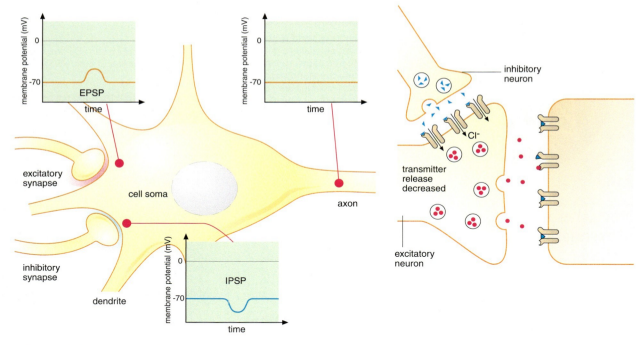

1 Excitatory and inhibitory synapse

2 Presynaptic inhibition

Neuronal switches — summation processes at synapses

A neuron in the central nervous system is not like a muscle fibre cell, which is only connected to one neuron. Each of the neurons of the central nevous system receives and processes information from up to 10,000 synapses. These neurons integrate the incoming signals and, depending on how much they are depolarized, send out action potentials of various frequencies.

Excitatory and inhibitory synapses

Two types of synapses connect nerve cells in the nervous system of humans and animals but these cannot be distinguished on the basis of their appearance; they are *excitatory* and *inhibitory synapses*. Neuromuscular junctions are always excitatory. In the central nervous system, excitatory and inhibitory synapses appear in the same ratio. Whereas excitatory synapses lead to a depolarisation of the postsynaptic membrane, inhibitory synapses cause a *hyperpolarisation*. During postsynaptic inhibition, the synaptic terminal releases transmitter opening the chloride ion channels of the postsynaptic membrane. Because of the influx of negatively charged chloride ions, the membrane potential decreases below the value of the resting potential and moves further away from the threshold that could initiate an action potential. The result is a so-called *inhibitory postsynaptic potential* (*IPSP*).

Another efficient way of controlling the conduction of impulses is by *presynaptic inhibition* (fig. 2). In this case, the inhibitory synapses are not located at the soma of the neuron but at the synaptic terminal of an excitatory synapse and cause a reduction in the release of the excitatory transmitter. This kind of inhibition is important for the regulation of muscle movement. It affects single synapses directly, e. g. the motor end plate in the muscle fibres of insects. Presynaptic inhibition also takes place in the spinal cord of vertebrates.

A chloride ion influx into the synaptic terminal lowers the amplitude of an arriving action potential. The low depolarisation reduces the calcium ion influx and thus the transmitter release is decreased. The EPSP of the conducting nerve cell remains below the threshold so that no action potential is triggered (fig. 2).

Synaptic terminal on the cell soma of a neuron

EPSP
excitatory postsynaptic potential

IPSP
inhibitory postsynaptic potential

inhibitory
Lat. *inhibere* = to restrain, to prevent

Spatial summation

Many impulses can reach the postsynaptic cell at the same time via synapses that are located at the soma of a conducting nerve cell. If several excitatory synapses that are located apart from each other are activated at the same time, it is possible to measure this in a higher amplitude of the EPSP on the cell soma (fig. 1). This is referred to as *spatial summation*. It can be explained by the greater number of released transmitter molecules. Even a single presynaptic action potential causes an EPSP on the postsynaptic cell via the transmitter. However, often the resulting electric field is not sufficient to trigger an action potential at the axon hillock. Single action potentials are then not conducted further. Only the summation of the depolarisations caused by many synapses on the membrane of the cell body leads to an EPSP above threshold at the axon hillock. The summation of many arriving action potentials results in one action potential in the conducting axon.

Temporal summation

If several action potentials reach a synapse via one presynaptic nerve cell within a period of only a few milliseconds, the generation of an action potential can be measured in the conducting axon (fig. 2). The postsynaptic potential disappears slowly. If a series of action potentials reaches the synapse, the subsequent potentials are added to the initial one. The fast conduction of a series of action potentials thus generates an EPSP whose amplitude is significantly larger than one generated from single action potentials. This triggers an action potential at the axon hillock. Spatial and *temporal summation* are based on the same process. The number of transmitter molecules is greater than in single impulses. An increased diffusion of sodium ions leads to a stronger EPSP. This is also true for IPSP, which conversely decrease the postsynaptic potential and thus hinder the generation of action potentials.

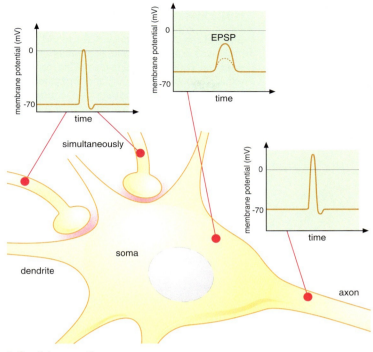

1 Spatial summation

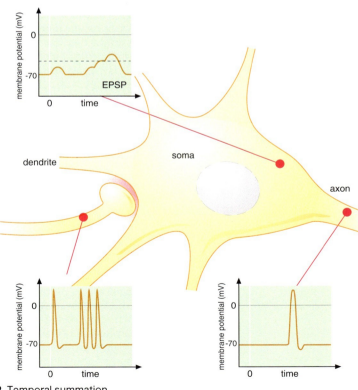

2 Temporal summation

Neurobiology

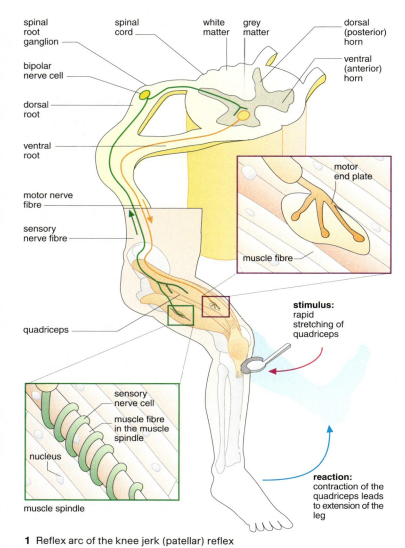

1 Reflex arc of the knee jerk (patellar) reflex

2 Crossed extensor reflex

Reflexes

Several simple genetically determined reactions are the same in all animals of a species. These reactions comprise, for example, breathing, coughing and closing the eyes when an object suddenly comes up close. These behavioural patterns are the normal response to a *stimulus* but without voluntary control and specific mental processing. They are called *reflexes*. Reflexes are based on a simple neuronal connection that allows a short reaction time. This is especially important in order to react to danger, e.g. the invasion of the trachea or eye by a foreign particle.

A careful strike just under the patella of a relaxed bent leg triggers an involuntary leg extension — the foot moves upwards (the so-called *knee jerk reflex*, fig. 1). Human skeletal muscles possess *muscle spindles*. These sensory cells within the muscle generate receptor potentials if the muscle is stretched or compressed. The *spinal cord* receives the generated impulse via fast-conducting *sensory nerves*. When the tendon is struck, the knee jerk reflex leads to the stretching of the quadriceps muscle. The sensory neurons are connected by synapses directly to *motor nerves*. Action potentials from them are transmitted by the motor end plates to the quadriceps and cause a contraction that results in the lower leg kicking upwards. Since only one central synapse is involved in the knee jerk reflex, it is referred to as *monosynaptic*. The brain only later receives information about the reaction.

The reflex arc as model

The basic principle of reflexes can be demonstrated as the *reflex arc* (fig. 25.1). It starts with a receptor on which the stimulus triggers an impulse. This is then conducted by sensory or afferent nerve tracts (neurons leading to the central nervous system) to the reflex centre. Here, the transfer to motor or efferent nerve tracts leading to the reacting organ (*effector*) takes place. The reaction times vary because the elements can be connected in several complicated ways. In humans, reflex centres are located in the spinal cord and in the brain, thus in the central nervous system. It is therefore possible to exert a voluntary influence on reflexes, e.g. to suppress coughing to some extent, even if an acute stimulus is present.

Neurobiology

Monosynaptic and polysynaptic reflexes

If you step on a pointed object, you involuntarily lift your foot and extend the other leg. This is the *crossed extensor reflex* (fig. 24.2) and the muscles of both legs participate in it. Additionally, a transfer of the sensory signal to several motor neurons is required. It is thus a *polysynaptic reflex*. Whereas in a *monosynaptic reflex*, a largely constant time between stimulation and reaction can be measured (knee jerk reflex about 30 ms), this time can vary in polysynaptic reflexes from 60 to 200 ms.

Reflexes can also be organised depending whether the same organ senses and reacts to the stimulus or whether different organs are involved. The knee jerk reflex and also the *blink reflex* belong to the former. Coughing belongs to the latter, because the stimulus is detected by sensory cells in the mucosa of the trachea and the resulting contractions occur in the diaphragm and the intercostal muscles. The muscle movements lead to a sudden increase of pressure in the lung. Thereby, a foreign particle can be forced out of the trachea.

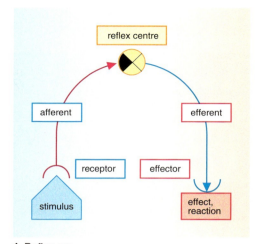

1 Reflex arc

Reflex
A stimulus-reaction process in which a specific stimulus in all individuals of a species triggers the same stereotypic, nervous-generated, involuntary reaction.

Tasks

① Trigger a knee jerk reflex. Pay attention to the sequence of reaction and conscious perception.
② Newborns show a grasp reflex that can be triggered by touching the inside of their hands. Classify this reflex and consider its biological significance.
③ Heart beat in humans is generated by continuous periodic signals of a muscle node (*sinus / sinoatrial node*). Compare this process with the course of a reflex.
④ Describe impulse conduction in a reflex arc by using fig. 2.

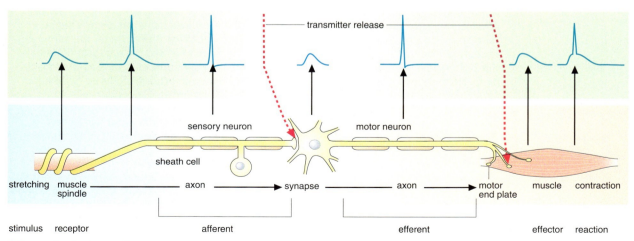

2 From stimulus to reaction

Neurobiology

3 Sensory organs

1 Human senses

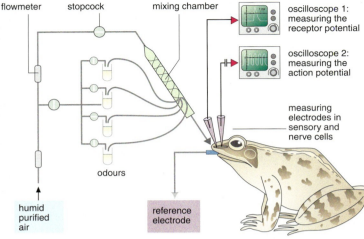

2 Testing the sense of smell

Receptors react to stimuli

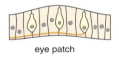

eye patch

pit eye

pin-hole eye

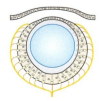

simple lens-containing eye

Is it possible to feel with the eyes? The answer seems to be yes, as being hit in the eye can prove: one sees "stars". Hereby, two things are special: there is a light sensation (stars) but a comparably large amount of energy is needed to accomplish this reaction.

Sensory organs translate the *stimuli* of the external world into the "language" of the nervous system. Neurons only react to electrical impulses. Thus, all types of stimuli have first to be translated into electrical potentials. This is the task of the receptor cells in each sensory organ.

According to the form of energy that they receive, *receptors* are classified as *chemo-*, *mechano-*, *photo-*, *thermo-* and *electro-receptors*. The traditional organization into the "five senses" is subjective and incomplete. Neither does it consider the enteroceptive senses in the inner organs such as the muscle spindles that measure muscle length, nor the CO_2 sensors in the arteries. A receptor can react to different stimuli but only one type of stimulus (the *adequate stimulus*) is able to trigger a response; moreover, it can do so even at an extremely low stimulus energy. The response of the receptor can be much stronger than the stimulus (signal amplification). In photoreceptors, it can reach about 20,000 times the energy of the received light.

Receptors are usually surrounded by supporting structures that help to process the stimulus (see margin). Together, they form the *sensory organ*. This can be simple, such as the pit eye of the snail, which only carries out basic functions, but it can also be highly complex and effective, such as the human lens-containing eye. Depending on the supporting structures, some eyes can only distinguish different light intensities, some can only register the direction of the light source, and some can recognize either simple or precise images of the environment.

Each receptor cell reacts to a stimulus by opening or closing specific ion channels. The stimulus causes an electric signal (*transduction*): the stronger the stimulus, the more changed the receptor potential. Primary sensory cells, such as the muscle spindle, react to a stimulus by generating action potentials. Secondary sensory cells, such as the photoreceptors in the eye, generate a receptor potential. Action potentials are only created in the connecting neurons.

Sensory nerves lead signals to the central nervous system. We perceive the sensation as light or movement depending on the location of the stimulated brain region.

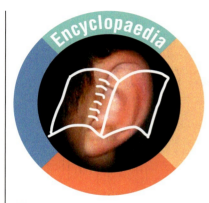

Human senses

Mechanoreceptors

Various types of *mechanoreceptors* can be found in the skin: *Merkel nerve endings* (1) are intensity sensors on which branches of myelinated axons end. They respond even when specific regions on the skin are pressed by only 10^{-6} m and they convey information about the depth of the impression and duration of the pressure stimulus. In hair-bearing skin, they occur as groups called *Merkel's discs*. *Meissner's corpuscles* (2) transmit information about the speed of the stimulus. In hair-bearing skin, *hair follicle receptors* (3) function as touch sensors. Their ends are wound around the base of the hair and react to any change in hair position. *Pacinian corpuscles* react rapidly and are thus able to detect vibration.

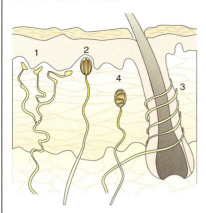

Sensory organs of the inner ear

Each part of the vestibule (utricle and saccule) contains a balance receptor. The cilia of secondary sensory cells extend into a gelatinous layer (*cupula*), which contains calcium carbonate crystals (*otolith*). When the head is moved, the cupula moves because of gravity and the cilia are shifted. The movement of the cilia triggers the opening of extension-gated ion channels and thus leads to depolarisation. The sensory cell secretes increasing numbers of transmitters and the frequency of the action potentials rises. A movement to the other side leads to a decrease in the action potential frequency. The *sensor for rotation*, which comprises three semi-circular canals of the *labyrinth*, is similar in architecture and function; each ampulla has sensory cells that reach into a gelatinous layer.

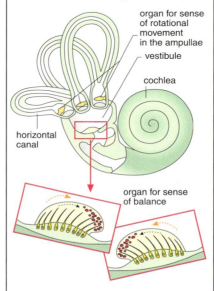

The *cochlea* comprises three parallel fluid-filled canals. The *basilar membrane* lies between the *median* and the *tympanic canals*. The *organ of Corti* is located on the basilar membrane. Sound waves transmitted from the ear tube to the oval window produce vibrations in the liquid of the vestibular canal. These pressure waves travel to the tip of the cochlea and, via the tympanic canal, back to the round window. The sound waves of a certain frequency cause a specific area of the basilar membrane to vibrate more than other areas: high frequencies lead to maximum vibrations at the basal end of the cochlea, whereas low frequencies (low-pitched sounds) lead to maximum vibrations at the apical end of the cochlea. These vibrations are recorded by sensory hair cells that are highly sensitive and can record rotations of less than 1 atom in diameter.

Chemical senses

The tongue is covered with *papillae* on the surface of which *taste buds* are located that are composed of bundles of 50 to 100 secondary sensory cells. Microvilli (finger-shaped processes of the sensory cells) terminate in the fluid-filled gustatory pore.

Ions reach the inside of the sensory cells by penetrating the membrane of the microvilli. Other substances bind to receptors on the outside. In both cases, the sensory cell is depolarized and excites nerve cells downstream. If the excitation is above the threshold, the nerve cells send action potentials to the brain.

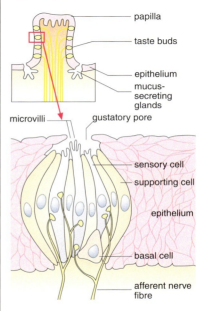

Odours have a similar mode of action. They penetrate the mucus layer of the nasal mucosa and accumulate on receptor cells on the dendrites of *olfactory sensory neurons*. Each cell carries several receptor molecules that respond to an odour of a certain type. They react to the binding of the correct substance by changing their membrane potentials.

Humans possess several thousand different types of receptor molecules. The building plan for each single molecule is determined within a gene; 3% of our genes code for these receptor molecules. Olfactory sensory neurons are primary receptors because they produce their own action potentials and send them via their axon to the brain.

The eye as a sensory system

Sensory cell (sensory receptor)
A cell that responds to environmental or internal stimuli by changing its potential or creating impulse.

When looking at an eye, we first notice the colourful *iris*, which has a hole for the black *pupil*. The iris is surrounded by the white *sclera*, which is covered by the conjunctiva, a transparent layer containing blood vessels. The *anterior chamber* lies between the transparent front part of the sclera, i.e. the *cornea* (more easily seen from the side), and the lens. This chamber is filled with a clear solution of salts secreted by the ciliary body (fig. 1). The *vitreous body* behind the lens supports the eye ball and contacts the retina. The *retinal pigment epithelium*, the *choroid* (the vascular layer) and the *sclera* compose the outer layers.

Millions of <u>sensory cells</u> are arranged in a precise pattern in the retina. Light passes into our eye through the pupil. Light entering the eye is refracted when passing the cornea and the lens. This produces a sharp picture of the environment on the retina. The flexible lens is surrounded by a ring of fibrous strands (*zonular fibres*) that are connected tightly to the *choroid* and the *sclera*. The intraocular pressure (the fluid pressure inside the eye) exerted on the sclera passively stretches the zonular fibres. This results in a flattening of the lens. The *refraction power* is thus small and the eye is *accommodated* to distant objects.

When looking at near objects, the circular *ciliary muscle* contracts. It is attached to the zonular fibres and its contraction counteracts the tension of the sclera. The pulling force of the zonular fibres on the lens decreases and, because of its flexibility, the lens becomes thicker. The lens is now more spherical and thus refracts the light to a greater extent. The increase in refraction power causes <u>accommodation</u> to near objects. An age-related decline occurs in the flexibility of the lens. It can no longer become as spherical and the refraction power decreases. People with age-related far-sightedness require spectacles with a converging lens for reading. Mammals and birds accommodate by changing the refraction power of the eye lens. Amphibians, fish and invertebrates, in contrast, change the distance between lens and retina.

When we perceive light, the direct adequate physical stimulus is represented by electromagnetic waves of wavelengths between 400 and 750 nm. Thus, we only perceive a small part of the whole electromagnetic spectrum. Radio waves, ultraviolet waves, heat rays and X-rays, for example, are invisible to us.

The visible world of animals can differ significantly from that of humans and is adapted to the living conditions of the respective animal. The *compound eyes* of insects, for example, have fundamentally different structures and functions. The spatial resolution is lower and the mosaic pictures are composed only of 300 to 3,000 single dots. Their temporal resolution is much greater. Whereas humans see a movie film with 25 pictures per second as a continuous flow, insects can distinguish more than 300 single pictures per second, which is excellent for detecting rapid movements. Many insects, e.g. bees, cannot see red but perceive ultraviolet light. Furthermore, they can detect the *polarisation* of light (the spatial orientation of the electromagnetic waves). This enables them to orient by using the position of the sun, even on cloudy days or when the sun is hidden from direct view.

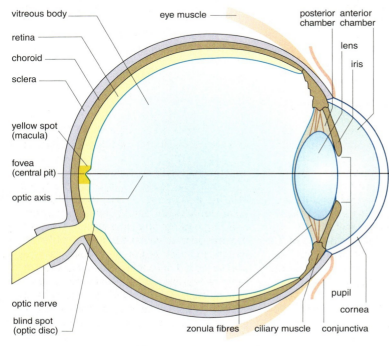

1 Horizontal section of the human eye

Compound eye

Neurobiology

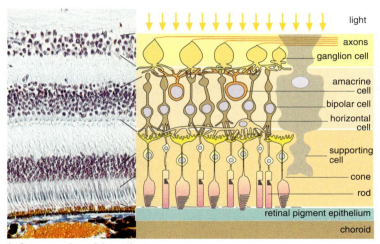

1 Cross section of the retina

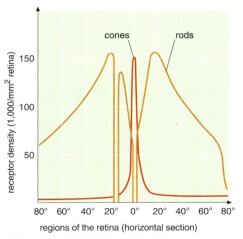

2 Receptor distribution in the retina

Structure of the retina

The sensory cells of the retina are located on the side opposite to the light. Two types of receptors can be observed by using a microscope: thin *rods* that distinguish between light and darkness, and skittle-shaped *cones* that are required for the perception of colour. The sensory cells have synaptic contacts on *bipolar cells* that, in turn, contact *ganglion cells*. The processes of the ganglion cells make up the *optic nerve*. The *horizontal* and *amacrine cells* are connected horizontally. Therefore, the stimulation of one sensory cell can influence several ganglion cells. However, there are many more sensory cells than ganglion cells, with 120 million rods and 6 million cones and, in contrast, only 1 million ganglion cells. Rods and cones are distributed differently across the retina (fig. 2). In the centre where the light of a focused spot reaches the retina, there are only cones. This part of the retina is called the *fovea* (*central pit*) and forms a small depression. Each cone is connected to a single ganglion cell, only here. This is the reason that we have to look at objects directly in order to see them clearly, as they are projected onto this spot. In general, more receptors are present in the centre than at the periphery. One part contains no receptors at all. This is called the *optic disc* or *blind spot*. Here, the optic nerve leaves the retina.

»info box«

The visual field

The visual field is the part of the environment that we can see when keeping our eyes stationary. There is a difference between the *monocular* (one eye) and *binocular* (two eyes) *visual field*. It is possible to measure the limits and losses in the visual field. The figure to the right shows the monocular visual field of the right eye.

When only light/dark vision is tested, the largest visual field is obtained. However, the visual field is more limited for colour vision. The figure also shows visual field loss caused by the blind spot, which we do not normally perceive in every day life.

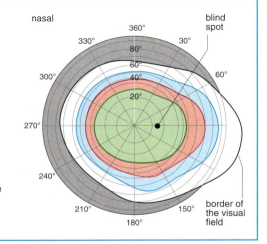

Neurobiology

The function of the retina

The *retina* of the human eye contains about 6 million cones for colour vision and 120 million rods for light/dark vision (fig. 1). Electromagnetic waves with a wavelength between 400 nm (violet) and 700 nm (red) cause the stimulation of the sensory cells. This is perceived as light. The structure of the two cell types is similar. The inner segment of each cell contains a nucleus, mitochondria and the synaptic terminal. The outer segment of the rods is made up of many stacked discs (about 1,000), which arise from folds of the outer cell membrane (fig. 2). This stack is surrounded completely by the membrane of the sensory cell. Molecules of visual purple (*rhodopsin*) are located in the membranes of these discs. As early as the last century, researchers proposed that the purple pigment of the frog retina is important for vision because the colour disappears when exposed to light and is restored in darkness.

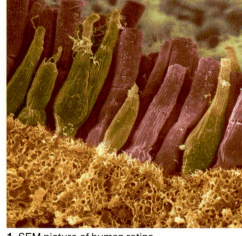

1 SEM picture of human retina

Rhodopsin molecules span across the membrane (fig. 2). *Opsin*, a protein, and *retinal*, produced from vitamin A in our diet, are parts of rhodopsin. Vitamin A deficiency can lead to impairment of vision, e.g. *night blindness*. The retinal molecule occurs in two forms: *cis* and *trans* (see margin). In darkness, the probability is 1 : 1,000 that the cis form changes to the trans form. During light exposure, the probability of the trans form occurring increases rapidly. Thus, both forms exist in a light-dependent equilibrium.

When measuring the receptor potential of light sensory cells, a surprising change in membrane potential is found compared with other sensory cells (fig. 31.2). Stimulation of olfactory (smell) sensory cells in the olfactory epithelium leads to a depolarisation. In light sensory cells, stimulation leads to a hyperpolarisation between -30 mV and -70 mV. Thus, the cell membrane is more depolarized without a stimulus. This can be explained by the hypothesis that a stimulus causes the closing and not the opening of sodium (Na^+) channels. Therefore, more sodium channels are opened in darkness than during light exposure.

The light-induced alteration of 11-cis retinal to all-trans retinal changes the conformation of opsin and triggers a signal cascade that leads to the closing of the sodium ion channels.

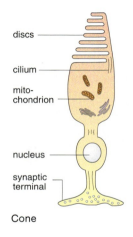

Cone

Light-dependent reaction of the photoreceptive pigments

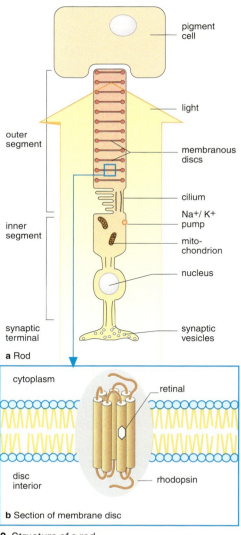

2 Structure of a rod

30 Neurobiology

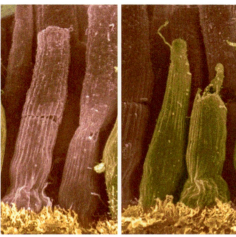

1 Rod (left) and cone (right)

tion of cones in the fovea is transferred directly without interconnected cells. Finally, the ganglion cells generate action potentials and send them via the fibres of the optic nerve to the brain.

Task

① Compare the states of the light sensory cell in figure 2 by making a table and explain the differences from a primary sensory cell.

At first rhodopsin is activated and reacts with transducin (a protein in the disc membrane of the rods). Activated transducin causes the enzymatic degradation of *second messenger* molecules (cGMP). The cascade leads to an amplification of the signal; one absorbed light quantum (photon) can lead to the cleavage of about 100,000 second messenger molecules. In darkness, these molecules bind to the rod membrane and keep the sodium ion channels open. The membrane is depolarized and transmitter is released in the synapses. Since these synapses are inhibitory, the activation of the connected neurons is inhibited.

The light-induced degradation of secondary messenger leads to the closing of the sodium ion channels and thus to hyperpolarisation. Therefore, less inhibitory transmitter is released into the synapses. The loss of inhibition leads to the activation of the connected neuron. Retinal is transferred back to the 11-cis retinal form by enzymes and the bleached photoreceptive pigments are regenerated.

Light sensory cells are secondary sensory cells. They do not produce action potentials. They transmit their stimulation to the *amacrine* and *bipolar cells* connected downstream. These cells integrate the potentials of the various sensory cells and pre-analyse them. Specific cells react specifically to contrasts and amplify them, whereas other cells react to specific shapes or movements. The bipolar cells pass their stimulation to the ganglion cells. Only the stimula-

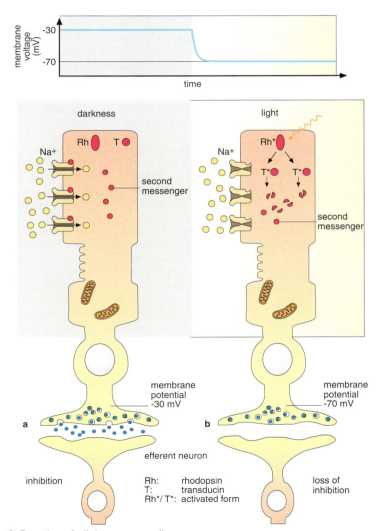

2 Function of a light sensory cell

Neurobiology 31

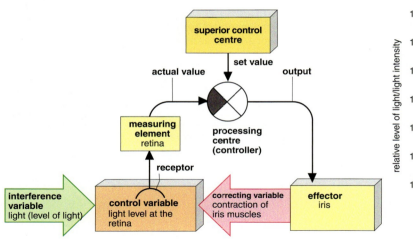

1 The pupil reflex as a control loop system

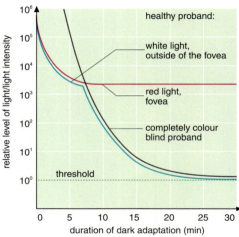

2 Dark adaptation curves

Adaptation: adjustment to light levels

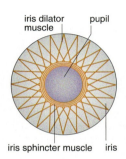

Capacity for iris adaptation

intensities

moonlight: 0.01 lx
summer day: 100,000 lx

contrast: 1:10,000,000

pupil size between 4 and 64 mm^2
control range of the pupil: 1:16

When we leave a highly illuminated room and go out into the night, we can hardly see anything. Weakly illuminated objects only become visible after a short period of time. The eye *adapts*, i.e. it adjusts its light sensitivity to the level of ambient light. Two processes are responsible for this.

In darkness, the pupil is opened widely. When more light reaches the eye, the pupil is constricted. The level of light reaching the retina has to be kept at an optimum. A model of the control loop (fig. 1) describes the processes as follows: a strong light stimulus causes a strong potential change in the retinal photoreceptors, which function as *receptors*. Ganglion cells send stimulations via the optic nerve to the brain, which acts as processing centre (*controller*). As a result, efferent nerves are stimulated and trigger the contraction of the iris muscles (*effectors*). The pupil becomes smaller and the amount of light reaching the eye is reduced. The *actual value* is adjusted according to the *set value*. In darkness, the opposite process takes place. The control loop counteracts a change in light exposure (*negative feedback*) and thus the *control variable* remains almost constant.

However, the size of the pupil can only be changed within fixed limits (see margin), which do thus not explain the complete *adaptation* capability of the eye. The main contribution is made by changes in the retina. Low light changes only small amounts of rhodopsin. The probability that an incoming light quantum meets an intact rhodopsin molecule is high. This can then cause a change in the potential of the sensory cell. Thus, even a few light quanta can be detected. After entering a dark room, only small amounts of 11-cis retinal are available and the main part is present as all-trans retinal. It takes some time until the light-sensitive form is completely regenerated. This is why dark adaptation takes time.

The opposite processes are faster. High levels of light that suddenly enter the eye are dazzling but also change a massive amount of photoreceptive pigment. Hereby, the probability that a light quantum meets a light-sensitive rhodopsin molecule and causes a potential change quickly decreases. The eye therefore adapts quickly to bright light.

The cones in the fovea adapt rapidly to darkness (fig. 2) but they also have a relatively high threshold. Rods adapt more slowly but they are much more sensitive to lower levels of light.

Apart from the changes in the amount of photoreceptive pigment, the number of light sensory cells to which one ganglion cell reacts is also changed. The number increases in darkness and decreases in bright light.

Neurobiology

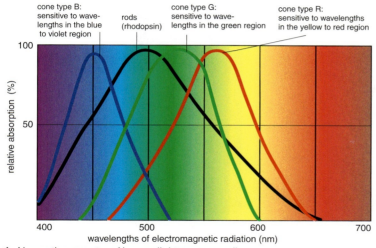

1 Absorption spectra of human light sensory cells

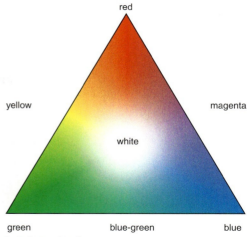

2 Colour triangle

Stimulus processing in the retina

Additive colour mixing

Subtractive colour mixing

We perceive light of a certain wavelength as a certain colour. The sensory cells in the retina respond to light of wavelengths between 400 and 700 nm, the spectral range between violet and red. We perceive a mixture of all these wavelengths as white light. The same result is obtained when mixing the spectral light of the colours red, green and blue. These *primary colours* mix additively. The mixture of two primary colours results in the complementary colour to the third primary colour. A mixture of the three primary colours yields white. Any colour can be produced by a suitable mixture of primary colours at an appropriate intensity. This principle is used, for example, in colour television.

In 1801, THOMAS YOUNG suggested the hypothesis that all colour impressions that we perceive consist of three primary colours. In 1852 HERMANN VON HELMHOLTZ concluded that three types of cones must therefore exist in the retina with each type being sensitive to light of a particular wavelength (*trichromatic colour theory*). The rhodopsin of the rods absorbs light of the entire visible spectrum (with a maximum at 500 nm, which is yellow-green). Cone pigments (e.g. iodopsin) have their absorption maximum at the blue (450 nm), green (530 nm) or yellow to red (570 nm) part of the spectrum. The three types of cones are stimulated differently depending on the light that reaches the eye. Light at 470 nm stimulates the blue receptor and the green receptor. We perceive the colour as blue-green. The same stimulation of all three types of cones is perceived as "white" or (in low light intensities) "grey". We can detect wavelength differences of 1 to 2 nm and thus can distinguish thousands of colour nuances.

However, we cannot distinguish the colours completely. Even in a dense forest, light seems white. A photograph of the forest floor shows the light conditions in an objective way: almost only green light reaches the floor through the leaves. The opponent colour theory of EWALD HERING (1874) suggests that colour impression is based on the proportion of stimulation of the opponent colour systems red-green, yellow-blue and black-white. Only if light of different colours reaches different places on the retina is the image perceived as colourful. Our visual system reacts to only a small extent to uniform colours but to a much greater extent to colour contrasts.

Is it possible to combine both theories? The stimulation of cones is analysed by the horizontal and amacrine cells and then transferred further to the ganglion cells. Hereby, the stimulation of the same colour receptors inhibit each other but stimulation of the different receptors enhance each other. The theory of HERING takes this level into consideration, whereas the theory of HELMHOLTZ and YOUNG considers the processes at the "cone level".

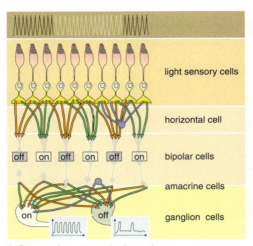

1 Processing stages in the retina

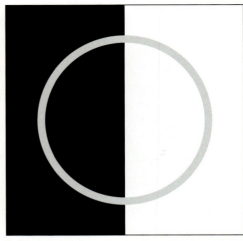

2 Grey contrasts

Receptive fields and contrast

The two halves of the circle in figure 2 are the same shade of grey. Despite this, the half of the circle on the dark background appears lighter than the other half on the light background. Thus we do not perceive absolute brightness values. Even in the retina, the stimulations are integrated by the rods and thus *contrasts* are increased. The biological significance of this is probably that, in twilight, enemies or food can be detected more easily on a background of a similar light intensity.

Neighbouring light sensory cells are connected by *horizontal cells* (fig. 1). In the dark, they release a transmitter that triggers stimulation (here hyperpolarisation) of the sensory cells. Light exposure stimulates the light sensory cells even more and they then inhibit the horizontal cells. Thus, the horizontal cells release less transmitter and stimulate the light sensory cells to a reduced extent. These, in turn, have a higher potential and no longer react strongly to light. This allows contrasts to be increased. (Model of *lateral inhibition*, see page 35). The ganglion cells at the next processing level are interconnected in a similar way by amacrine cells. They are also programmed such that, when a cell is stimulated, it inhibits its neighbouring cell.

The brain does not receive stimulations from the cones or rods; it receives the action potentials of the ganglion cells. Each light sensory cell is connected to a single ganglion cell only at the fovea. In all other regions, a ganglion cell integrates the stimulations of many bipolar cells and, thus, the stimulations of up to several thousand light sensory cells. They form the *receptive field* of this ganglion cell.

Each bipolar cell is in contact with several light sensory cells (fig. 1), which can stimulate (ON type bipolar cell) or inhibit (OFF type bipolar cell) their bipolar cell when exposed to light. Several bipolar cells are in turn connected in a stimulatory or inhibitory way to a ganglion cell, which receives and processes information from many light sensory cells. Hereby, the stimulations of the centre of the receptive field work against the stimulations from its periphery. Strong light in the centre of the receptive field and, at the same time, weak light at its periphery can stimulate a ganglion cell and trigger it to produce more action potentials (ON ganglion cell). It can also inhibit the ganglion cell and reduce the number of action potentials produced (OFF ganglion cell). An even light exposure at both the centre and the periphery causes stimulation and inhibition of same strength and thus does not change the response of the ganglion cell. It is therefore not perceived by us.

The processing of stimulations in the eye is therefore adjusted to recognize contrasts and changes. This includes differences in brightness, colour contrasts and pattern changes over time (movements).

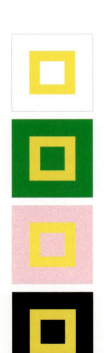

Colour contrast according to Johannes Itten (1888 — 1967). The colour impression is influenced by the surrounding colour.

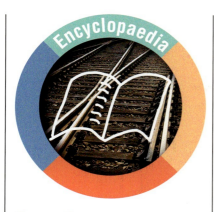

Perception

Optical illusions show how the brain analyses optical information.

Ambiguous illusions

Experience, interest, expectation or also the mood of the moment can cause bias and stimuli are perceived selectively. The viewer preferentially perceives what he expects to see. This decides what the observer initally sees in ambiguous images.

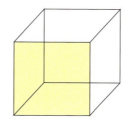

The spatial perception of an image can also be ambiguous. The colourful side of the cube can be perceived as either the front or the back. The cube "jumps" if it is watched for a long while.

Impossible objects

Because of its experience in a 3-dimensional world, the brain always tries to produce a 3-dimensional image from a 2-dimensional picture. Any angle is interpreted, if possible, as a right angle that is skewed in perspective. It is not possible to resist this, even if the interpretation of such bodies cannot exist in reality. The otherwise helpful tool of natural orientation thus fails.

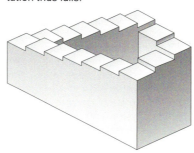

Geometric illusions

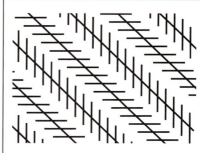

When we look at the figure above, we are misguided by an angle illusion: parallel straight lines seem to converge; the acute angles are overestimated because the straight lines are crossed by many of short parallel lines. The illusion disappears when the line of vision is parallel to the straight lines (from the corner of the image).

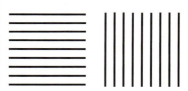

Our eyes are used to eye movement from left to right (or right to left in Arabic) because we read and are thus faster in this direction than in movements from top to bottom. Thus, we tend to overestimate vertical lines so that height is overestimated. This effect also applies when wearing striped clothes.

Model of lateral inhibition to increase contrast

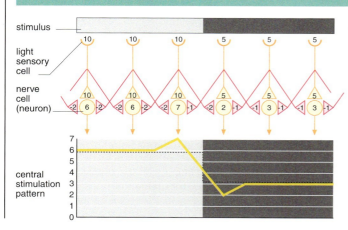

A simplified model explains how lateral inhibition increases contrasts. In the light area, the light intensity is double (10) that in the less light areas (5). Horizontal cells inhibit the amount of stimulation by 1/5. Based on this model, each sensory cell in the light area is inhibited by two horizontal cells and thus reacts with a stimulation of $10 - 2 - 2 = 6$.

This functions similarly in the less light area: $5 - 1 - 1 = 3$.
At the brightness boundary, the situation is different. Here, each sensory cell is influenced by one strongly and one weakly reacting horizontal cell. Its stimulation is decreased by 3. It passes on a stronger stimulation of 7 (light) or a weaker stimulation of 2 (dark).

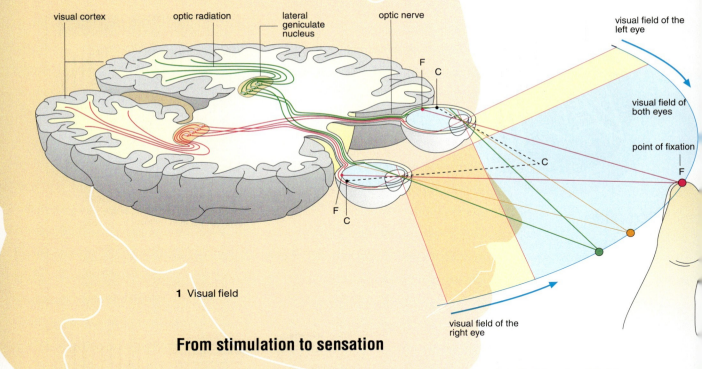

1 Visual field

Visual field
The area within which an object can be seen without moving the eyes.

From stimulation to sensation

A red car is on a street that we want to cross. The various optical stimulations in such situations trigger electrical impulses in the light sensory cells in the retina. These are transported via interconnected neurons to the brain. About 1 million neurons form the optic nerve, which leads from the eye to the brain. The optic nerves of both eyes meet in the optic chiasm. Here the nerve fibres cross partially to the other side of the brain. They are distributed in such a way that the nerve fibres that have their origin in the right half of the retina of each eye end up on the right side of the brain, and similarly for the left half. As a result, the visual information from each half of the visual field seen with both eyes reaches only one half of the brain (fig. 1).

Signal processing

The nerve fibres of the optic nerve end in the diencephalon. The signals are processed here and forwarded to the visual cortex, a region in the cerebral cortex. The neurons in the cerebral cortex have a differing selectivity for different visual stimuli within the visual field and form three processing systems: one each for the perception of movements, form and colour.

The first system provides washed-out pictures with poor contrast and without colours. All moving objects are preferentially distinguished. Here, position and movement are analyzed. The second system delivers clear forms of parts of the whole picture. Here, the object is identified. The third system provides images that are not sharp but that are colourful in the middle region.

Depth perception (the visual ability to perceive in 3-dimensions) is produced in the first and second system by using slightly different images on the retina. The difference is attributable to the distance between the eyes. When we concentrate on an object, both eyes are turned in such a way

»info box«

Disturbances in the perception of motion

As early as the end of the 19th century, SIGMUND FREUD observed that some patients were not able to recognize visual attributes. He concluded that the defects were not in the eyes but in the brain.

One female patient lost the perception of movements after a stroke but she could still recognize colours and forms. She had problems pouring tea into a cup because she could not see the fluid move. She saw it as a "frozen liquid". Additionally, difficulties occurred because she could not see the fluid level rising in the cup. Problems also arose on the street. She could see cars but she could not determine their speed because the cars were first there and then all of a sudden here without her seeing the cars move. This led to difficulties in crossing the street. She had to learn to include other factors for orientation, e.g. the increasing noise of the cars.

36 Neurobiology

that the picture of the object is in the fovea of both eyes. Objects located in front of or behind this fixation plane produce images next to the fovea. Objects in front of the fixation plane (C) are produced in the outer part and objects that are further away in the nasal part of the retina.

Seeing with eye and brain?

Perception of the environment is the result of the signal processing of the three systems. Information that produces the most attention governs the perception, while other objects are ignored. This is comparable to illuminating the surroundings with a spotlight: only some parts are illuminated, while the rest remains in darkness. When crossing a street, the movement of the car draws our attention. This is evaluated in the brain, which makes us wait at the side of the street (fig. 1).

During the visual perception of our environment, only 20 % of the impulses processed in the cerebral cortex originate from the light sensory cells of in the retina. Most impulses come from other regions of the brain. The stimulus pattern of the environment becomes an impulse pattern in the brain. Together with saved experiences, the impulse pattern leads to a recognition of the surroundings, called *cognition*.

1 Blue dots or a picture?

This is obvious in figure 1. When looking at the picture, only random blue dots can be seen. When looking at the picture a second time, with the additional information that here a horse and a rider can be seen, the whole picture becomes visible. The additional information changes the evaluation of the signals from the eyes to the brain and a new perception is produced in our conscience.

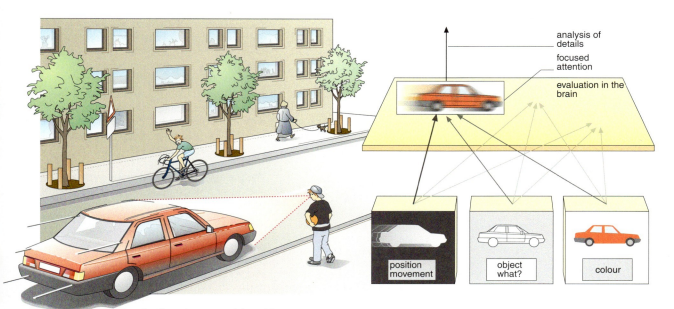

2 Three channels of evaluation when perceiving objects

Neurobiology 37

4 Structure and function of the nervous system

Human nervous system

The human nervous system is very efficient. The number of neurons forming it is estimated to be about 10^{11}. The brain and spinal cord together are called the *central nervous system* (CNS).

The *spinal cord* has two functions: It is the central connecting element between the brain and the *peripheral nervous system* (fig. 1), which is distributed throughout the body. It is also the independent "switching point" of sensory to motor neurons (see *reflex arc*). *Interneurons* often connect the sensory and motor neurons. The connecting and converting functions are allocated to anatomically different regions in the spinal cord. The connecting tracts are located in the outer white matter, whereas converting occurs in the inner grey matter. A spinal nerve with its ventral and dorsal roots originates to the left and to the right between two vertebrae. Sensory neurons direct impulses from the sensory cells to the spinal cord (into the *posterior horns*) via the *dorsal roots*. The body of each of these neurons lies outside the spinal cord in the *spinal ganglia*. The *ventral roots* (from the *anterior horns*) send motor nerve fibres to the muscles of the upper body. The cell bodies of the motor neurons are located in the grey matter of the spinal cord. In addition, the grey matter is composed of a network of dendrites and mainly short unmyelinated axons. The two roots fuse forming a mixed nerve just after the spinal ganglion. Humans have in total 31 pairs of spinal nerves.

The autonomic nervous system

The spinal nerves also innervate the inner organs. Most of the effects of this part of the nervous system cannot be controlled consciously. Therefore, it is called *autonomic* or *vegetative*. Its task is to control the inner milieu (homeostasis). The autonomic nervous system is divided into two functional subsystems: the parasympathetic and sympathetic nervous systems. Depending on need, the various organs can be activated or inhibited. The autonomic nervous system hereby works closely together with the somatic part of the nervous system, which connects the skeletal muscles to the spinal cord.

For example, when a dangerous object suddenly approaches, an individual is quickly sent into a state of highest efficiency. All aspects of the body that help to achieve this state are initiated and coordinated by the sympathetic nervous system (fig. 39. 1). The blood circulation is increased so that, for example, muscles receive an optimal amount of oxygen and nutrients. All processes that are unnecessary for overcoming the danger are decreased.

The parasympathetic nervous system often acts in an antagonistic way to the sympathetic nervous system. Both systems frequently innervate the same organs and control vital functions of blood circulation, digestion, defaecation, metabolism, secretion, body temperature and reproduction. The processes during digestion are mainly carried out when the body is at rest. They are initiated or facilitated by signals of the parasympathetic nervous system. Further-

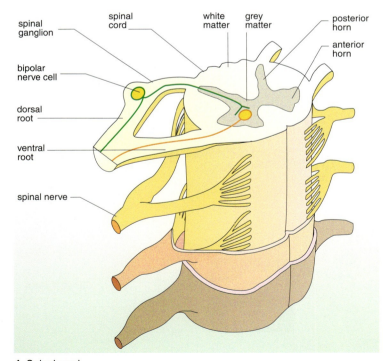

1 Spinal cord

more, together with the brain, the parasympathetic and sympathetic nervous system influence our emotional mood. If we are alarmed by a dangerous situation, the *limbic system* of the brain (see page 40) triggers the adrenal glands to release high concentrations of the hormones *adrenalin* (*epinephrine* in the US) and *noradrenalin* (*norepinephrine* in the US) into the blood circulation. Feelings such as fear, anger and happiness are accompanied by typical body reactions that are initiated by the autonomic nervous system. To what extent feelings govern a person depends on the superior parts of the brain. Noradrenalin is also produced as a transmitter in the synaptic terminals of the neurons in the sympathetic system.

It seems to have its effect here independently of mood changes. The transmitter of the parasympathetic nerve cells is *acetylcholine*.

Task

① In stressful situations, the sympathetic system supports efficiency. Diseases of civilisation, e. g. increased risk for infarction, enlargement of the adrenal glands or disturbed sexual behaviour, can be traced back to continuous stress. Explain the connections (fig. 1).

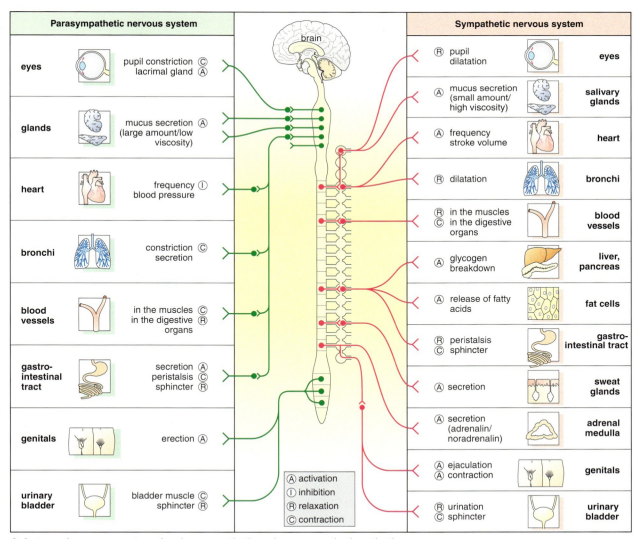

1 Autonomic nervous system showing sympathetic and parasympathetic activation

Neurobiology

Structure and function of the human brain

1. Cerebrum or telencephalon

The human *cerebrum* makes up about 80% of the brain volume. It consists of the *cerebral cortex*, which in humans has an enormously enlarged surface, because of the presence of furrows (sulcus, plural: sulci), and billions of neurons. The white matter lies beneath it and is a neuronal network. It enables the areas of the cerebral cortex to communicate with each other. The cerebral cortex has two halves: the *cerebral hemispheres*. They do not perform equivalent functions in humans. In most people, the right hemisphere is responsible for the non-language integrative processes of the whole information content, whereas the left hemisphere is for language-based processes and the analysis of details.

The cerebral cortex can be divided into different functional areas. Here, the incoming information from the sensory organs is combined and compared with previously saved information (experiences). The parts close to the forehead are responsible for initiative, the planning of actions, social behaviour and the processing of the content of the short-term memory. The regions for language and for the control of the mouth muscles are located here and the contents of the episodic memory and the declarative memory (facts) are stored in this region. In the parts on the sides of the brain, language and sounds are perceived and processed. These part are essential for storing and consolidating information. The processing of visual stimulation occurs in the brain near the back part of the skull.

2. Corpus callosum

This part of the brain mainly consists of nerve fibres that connect the two hemispheres. When the corpus callosum is surgically cut in two, information exchange between the two hemispheres is lost. If an object enters the left visual field of such a patient in such a way that the information of the retina reaches only the right hemisphere, the patient is not able to name the object, because the centre for language is located in the left hemisphere.

40   Neurobiology

3–6 Interbrain

This region is composed of the *thalamus* with the attached *epiphysis* (pineal gland), with the *hypothalamus* and *hypophysis* lying below.

The upper **thalamus (3)** is an important relay centre for all afferent tracts of the sensory organs (except the sense of smell) passing them on to the cerebrum. Here, the impulses are processed before they gain access to the cerebrum and thus come to conscious notice. This is why the thalamus is also called the door to the cerebral cortex.

4 Hypothalamus

Located below the thalamus is the **hypothalamus (4)**, which is the control centre of the autonomic nervous system and the hormone system. Its central function is to control the biorhythm of the body (see page 74). The hypothalamus secretes *releasing hormones* that affect the hypophysis underneath it. Motivational and emotional behavioural patterns are also regulated here.

5 Hypophysis

The **hypophysis (5)** is the size of a cherry and, controlled by the hypothalamus, releases hormones that are called *stimulating hormones*. It is involved in the control of body functions such as heat balance or sexuality. The hormones also control the maturation of the ovum, pregnancy, growth, water balance and basic metabolic rate. In the case of egg maturation, one of the specific hormones of the hypophysis is follicle-stimulating hormone (FSH).

6 Epiphysis

The **epiphysis (6)**, which is generally called the *pineal gland*, is a light-sensitive organ in some vertebrate classes whose hormones are involved in colour changes of the skin. In mammals, it secretes the hormone melatonin. Melatonin regulates functions that are related to light and seasonable changes, e.g. the sleep rhythm.

11 Medulla oblongata

The *medulla oblongata* (elongation of the spinal cord) is an evolutionary old part of the brain. It is the centre for vital reflexes such as salivation, swallowing, vomiting, coughing and sneezing and is the autonomic centre for breathing, heartbeat and blood pressure. If these vital functions are disturbed, for example, because of a broken neck, death occurs immediately. The medulla oblongata together with pons and midbrain are called the *brainstem*.

7 Hippocampus

This is located, as part of the cerebral cortex, in the inner margin of the temporal lobe and has many different functions such as attention, short-term memory, social behaviour and the processing of fear and spatial relationships. It is also important for the perception of our own body. Its name is based on its sea-horse-like shape.

9 Cerebellum

The cerebellum controls motor and balance functions. Impulses that are important for body posture and balance travel from and to the motor centres of the cerebral cortex via the cerebellum, which passes its impulses either directly to the motor centres of the cerebral cortex or to the body via tracts of white matter in the spinal cord. Here, learned procedures and coordination, such as driving a car or riding a bike, are stored. Loss of function in the cerebellum does not lead to the loss of certain motion procedures but to a loss of the coordination of movements.

8 Midbrain

This region forwards impulses from eye, ear and surface receptors to other centres of the brain. It is additionally responsible for fast orientation in the optical field. Here, movement is perceived, as is the appreciation of "where". What we see is however first processed in the cerebral cortex. Auditory and pain perception are also interconnected in the midbrain.

10 Pons

The pons connects the hemispheres of the cerebellum and passes impulses from the cerebral hemispheres to the cerebellum. It is partly responsible for sleep, waking and motor functions. It is active during dreaming.

12 Limbic system

This is a collective name for parts of the cerebrum and parts of the midbrain. The relevant parts of the limbic system are: the hippocampus (7) and amygdala (12). It is essential for transferring information into long-term memory. Because of its involvement in emotional evaluation, it is highly important for learning processes and the retrieval of information from the cerebral cortex.

Neurobiology

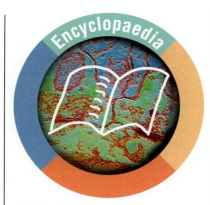

Methods in brain research

Even in ancient Greek culture, anatomical descriptions of the human brain existed. Mental powers such as reason and memory were thought to be located in the fluid-filled spaces of the brain, the *ventricles*. Not until the 19th century did researchers manage to assign functions to specific regions of the human brain.

The first research methods were, for example, the exact observation of neurological deficits and the localization of the injured parts of the brain. For this purpose, the brains of the patients were dissected after death and the anatomical changes were examined.

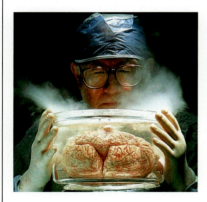

In 1909, the researcher HARVEY CUSHING introduced a method that made it possible to stimulate certain regions of the cerebral cortex in the living human brain. The cerebral cortex has no pain receptors and thus the operation and stimulation can be performed on a fully conscious patient. Different regions of the cerebral cortex were stimulated with weak electric currents. The patients felt faint tingling sensations on certain parts of the body surface and could make statements as to perceived sensory stimuli that were not actually present.

Modern methods enable us to examine the brain without opening the skullcap. Two methods are *electroencephalography* (EEG) and the measurement of the perfusion of blood through the brain ("*cerebral blood flow*"). Using these methods, we can assign certain activities or functions to individual brain regions. Images of the brain can be obtained by tomography without opening the skull. The use of this has made it possible to save and evaluate enormous amounts of measured data.

Electroencephalography

This word is derived from Classical Greek: *enkephalos* = brain, *graphein* = to write. The method is based on the observation that electric potentials can be measured not only on the surface of the brain, but also on the closed skullcap. The potentials are characteristically wave-shaped (see figure). In order to measure the potentials button-shaped electrodes are placed on the scalp.

A second electrode is placed at a certain distance away, e.g. on the earlobe, so that the voltage fluctuations between these electrodes are measured. The potential differences are extremely weak, lying in the range of microvolts, and thus have to be amplified. The amplified measured values are continuously recorded and subsequently evaluated.

Animal experiments have shown that these potential changes are *excitatory postsynaptic potentials* (EPSP). The electrodes on the scalp do not measure the activity of individual synapses, but rather the EPSP of about 1 million neurons.

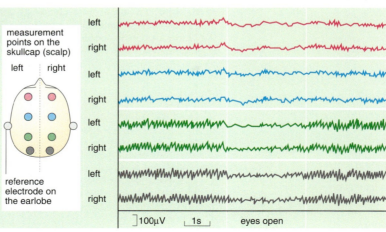

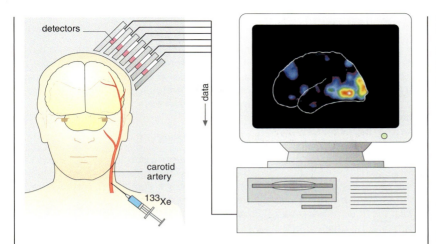

Important areas of application are the diagnosis of seizure disorders (e.g. *epilepsy*), the certain determination of *brain death* or the monitoring of the depth of anaesthesia. Analysis of the frequency and the amplitude allow conclusions to be made about the examined person. During an epileptic seizure, the amplitudes of the waves increase and the frequencies increase significantly. In the case of brain death, all eletric activities cease.

This procedure is important for *sleep research* because the various characteristic waves provide information about the depth of sleep during the different sleep stages. Thus, some sleeping disorders can be assigned to certain types of waves.

Blood perfusion of the brain

The energy metabolism of the brain is extremely high; 20 % of the inhaled oxygen is needed in the brain, even though the mass of the brain represents only 2 % of the body mass. The enormous neuronal network can only function if oxygen and glucose are supplied non-stop by the blood flow.

Activity in the brain leads to increased energy metabolism and thus increased oxygen uptake within seconds within the active region. These changes of the energy metabolism in the neuronal cells can be measured by using small harmless amounts of radioactive noble gases, e. g. radioactive xenon gas is injected into the carotid artery or is inhaled (see figure left). Because of the intense blood perfusion, the gas is quickly distributed in the brain tissue. It is also flushed out fast from the tissue by the blood stream and then exhaled. Depending on the level of perfusion, removal from the brain tissue occurs at different rates. The differences in radioactivity are measured with detectors placed on the sides of the head and are processed by a computer. Individual intensities of radioactivity are represented by different colours on the screen and then evaluated.

In PET (*positron emission tomography*), glucose that is marked in a positron emitter is used instead of the radioactive noble gas. The positrons fuse with the electrons and release rays that can be measured. The active regions of the brain are those that are rapidly using glucose.

Tomography

By using *tomography*, the brain of a patient can be shown as images in different planes. The layer thickness of the individual planes is 5 to 10 mm. Such images are important for medical investigations, especially when examining the patient for brain tumours. Ultrasound examinations that are often employed clinically are not useful in the skull area because of the large amount of bone mass. *Magnetic resonance imaging* (MRI) makes it possible to save the data in a computer and thus to evaluate images from the skull area.

The measurement involves the physical properties of hydrogen atoms, which are present in the water of the tissue and in all molecules in the brain. The atomic nuclei of the hydrogen atoms produce a magnetic field. During measurement, a magnetic field is produced that penetrates the skull. The hydrogen nuclei align like the needle of a compass with the applied magnetic field. When the external magnetic field is turned off, the hydrogen atoms swing back into their original positions and release electromagnetic rays in the range of radio waves. These are detected by receivers outside the body and are saved in the computer for processing. Details of down to 1 mm can be evaluated in the images produced.

Neurobiology 43

Learning: storage – retrieval

Learning makes us think mainly of classes in school. However, learning processes occur in our brain all the time. We remember routes, the taste of food or people's faces.

Stimuli from the environment are converted into impulses by sensory cells and forwarded to the brain by neurons. The incoming amount of information depends on the type of stimulus: olfactory stimuli (taste) can contain about 20 bits per second, whereas visual stimuli in contrast can amount to about 10 million bits. This flood of information could eventually block the brain. Therefore, the amount of data that finally reaches the long-term memory and is stored there is reduced by the memory systems involved in processing the information before it actually reaches the long-term memory.

Sensory memory

The *sensory memory* only stores received information within the sensory organs. Incoming stimuli are converted into impulses and are forwarded within half a second to several seconds for further processing.

Short-term memory

The *short-term memory*, whose name comes from the short period of time for which information is stored here, can retain information for about 10 seconds to minutes. However, the capacity of the short-term memory is restricted to a few units of information. Here, they can be connected to previously stored information. Depending on personal interests and mood, information is evaluated and receives a meaning based on previously stored information. New information is forwarded, especially if associations to it are already present. If the information in the short-term memory is not associated to the content of the long-term memory, it is lost forever. The short-term memory is located in the cerebral cortex. If the incoming information can be associated with previously stored information, it is stored in the cerebral cortex. In this process, the left hemisphere of the *cerebral cortex* is responsible for information concerning neutral knowledge and the right hemisphere for experiences and emotional information. In order to pass the information that is stored in the short-term memory to the long-term memory, it has to be consolidated, which means that it has to be connected to previously made associations or that it must be repeated so often that a threshold to the long-term memory is overcome.

Long-term memory

New information is permanently stored in the *long-term memory*. If we cannot remember information that is stored here, this could be because it is overwritten by other information or the process of retrieval is inhibited. The *limbic system* with the *amygdala* is important for the memory-storing process. Here, the impulses are tested for biological and social importance and compared with previously stored information. At the same time, the impulses are connected to emotions. The process of learning and the retrieval of stored information are connected to emotions. People who have a non-functional amygdala can neither recognize emotions (facial expression of happiness or grief) nor integrate facts. Emotions are extremely important for the evaluation of incoming stimuli.

Memory systems

Based on the findings of brain research, long-term memory is divided into an unconscious storage of information (procedural memory and priming) and a conscious storage of information, which is called declarative memory and includes semantic, episodic and perceptual memory (fig. 45.1). Knowledge is not stored as a whole or in clearly defined "boxes" but is stored all over the brain. Cornerstones of knowledge are stored according to the preference of each individual and, depending on their quality, this happens at different locations in the brain. Colour impressions are stored in locations different from those for information about the shape or type of material of an object or about smells and tunes. When memorizing, the brain remixes the learned information from the cornerstones. Every memory that we retreive has therefore been changed, based on our personal (experience), and has thus been interpreted. This is why memories are not identical to the original object or occurrence.

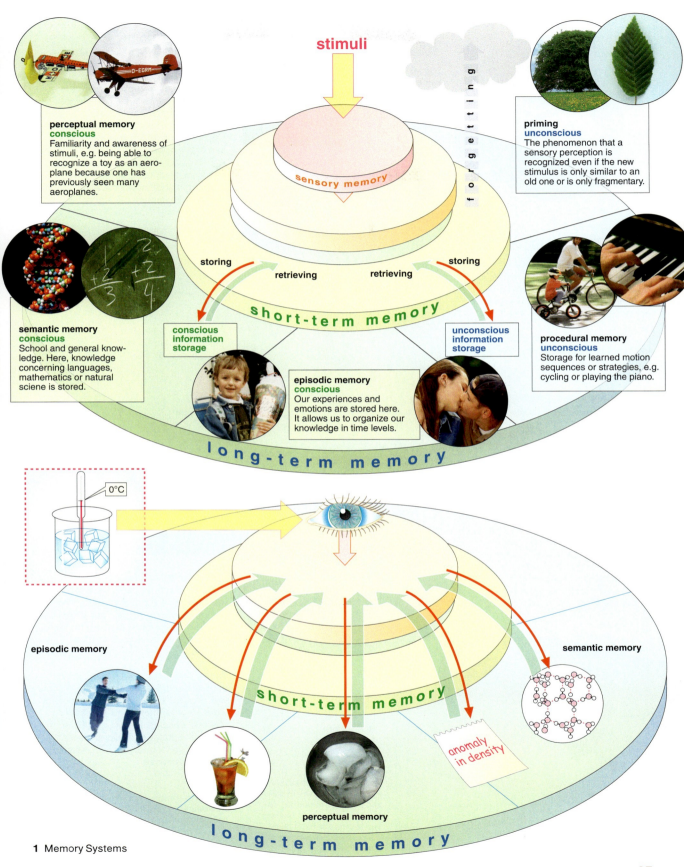

1 Memory Systems

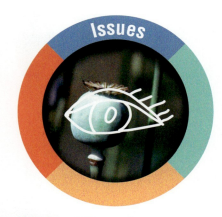

Issues

Psychoactive substances

"Up one minute, down the next". In every day life, our mood lies somewhere in between. We only realize what mood we are in if someone asks us; we do not normally think about it.

Our *mood* is the result of processing sensations, signals from the nervous system and hormones.

Psychoactive substances in the body

Running a marathon, skydiving and freeclimbing have one thing in common for the participants: overwhelming feelings of happiness can arise even though the situations are, for most people, extremely exhausting or scary.

The reasons for this are "the body's own drugs", which are substances that influence consciousness. These endorphins also reduce the physical experience of pain and exhaustion. They cause feelings of happiness. They were discovered during research on the effect of opiates.

The discovery of opiate receptors

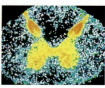

Opiate receptors, which were made visible by using radioactive opiate molecules, in the spinal cord.

The greatest receptor density is where the sensory fibres from the spinal cord and the grey matter meet (orange). Here, the so called *pain fibres* have synaptic contacts to neurons whose axons extend to the brain.

Opiate receptor molecules have been found on the pain fibres. Opiates lower the transmitter release in the synaptic terminals of these pain fibres. The pain threshold is thus raised. The pain-relieving effect of *opium*, an extract of the sap of the opium poppy has been known for a long time. Like the synthetically changed product obtained from it (heroin), it additionally causes euphoric states. It influences the signal transmission of certain synapses in the brain.

Explain the connection.

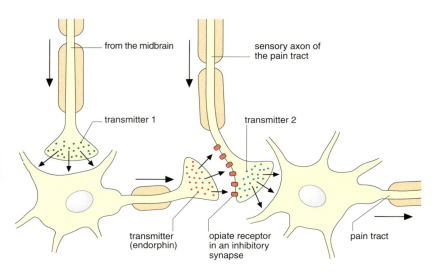

Drugs — a term with two meanings

Psychopharmaca
Medicine (pharmacon) that can relieve or heal psychological disorders.

Everything is poison — there is poison in everything. Only the dose makes a thing not a poison.
PARACELSUS (1493 – 1541)

Not only athletic maximum performance but also daydreams and pleasant memories can trigger comfortable feelings and feelings of happiness. Why can sports become an addiction?

Why are there no drugs in a drugstore? Originally parts of plants used for medications were preserved by drying. They were called drugs. They were used as a tea or as a moist poultice to put on wounds or were taken to combat diseases.

Today, drugs might be psychoactive substances that are sold legally — some with an age restriction (e.g. alcohol and nicotine) — or that are forbidden, e.g. hashish, cocaine, heroin, or LSD. Physical damage and addiction are possible consequences of consuming these substances.

In most cultures, the consumption of drugs was bound to religious procedures. Find examples and explain the connection.

Neurobiology

A second-hand sense of well-being

It is no coincidence that the original form of the brown soft drink has Coca in its name. It originally contained cocaine as stimulant; today, it only contains caffeine from the seeds of the cola tree. Cough syrup originally contained cocaine to dampen the cough reflex.

Cocaine inhibits the re-uptake of the transmitter dopamine into the synaptic knob. The dopamine concentration in the synaptic cleft is thus increased.

Altered perception, changed feelings. How can you explain these effects of cocaine?

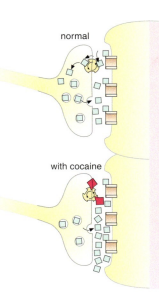

normal

with cocaine

The dangers of cocaine addiction

After a while, cocaine consumers can no longer have normal feelings of happiness without the drug to amplify it.

Consequence: severe depression, psychological addiction

Influencing the amplifier system leads to an overload of the brain.

Consequence: hallucinations

Possible consequences of the drug for the organism: Heart rhythm disturbance, heart infarction, stroke, epileptic seizures

Happy as the cat that got the cream

The existence of endorphins provides a hint about the biological importance of feeling good. To make it simple: lust and avoiding frustration are essential driving forces and are not restricted to human behaviour.

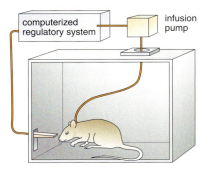

Interviewing drug addicts

I take drugs ...

because drugs improve
my mood/lift my spirit _____ 6.9%

because they make me
feel happy _____ 5.5%

because I lose my
inhibitions _____ 4.9%

What do you think about these motives? What risks are taken by the addict? Are they worth the risks?

Ecstasy — fun for parties?

In 1913, a substance was developed as a slimming agent but it was never released onto the market in Germany. It is called *MDMA* and, today, is the drug called *ecstasy*. It is illegally produced as pills - in different doses and with different additives — and is sold quite cheaply almost everywhere. Apart from feeling high, consumers ignore important signals of the body, such as exhaustion and thirst.

Dangers due to consumption? Gateway drug? Long-term damage? Important questions that we should try to answer.

If mice, rats or apes are given access to cocaine, they consume the drugs continuously during the experiment. Other usually positive activities such as eating, drinking and mating are neglected.

limbic system

Addiction

Addictive behaviour is part of our emotional repertoire and thus our personality, e.g. yearning, jealousy, gambling and greed. Ultimately, every human interest can be addictive, such as watching TV, work addiction and compulsion disorders involving excessive cleaning or addiction to collecting items. In all drug addictions (dependency on psychoactive substances), the following phenomena can be seen:

— development of tolerance
— physical dependency
— psychic dependencies

What do these terms mean? How does one become addicted? How can one overcome it? What should you do if your friends become or are addicted?

Medication that relieves anxiety

Some people suffer from anxiety without having an appropriate reason. Such a disease can be treated with psychotherapy or in severe cases with medication. *Barbiturates* have a calming effect (sedate) and are also used as sleeping pills. Someone who is tired often feels less anxious. Some drugs reduce fear and are sleep-inducing at higher doses.

However, addiction to these drugs is a danger. The effect is based on their binding to receptors in the limbic system and the cerebral cortex, instead of to transmitters.

Pathological fear is rare. What scares us? Are other options available, rather than turning immediately to pills?

5 Hormones

Hormones
These are chemical signalling substances that are produced by specific cells (often glands) in small amounts and that are then transported to effector cells via the blood stream.

Transmitter hierarchy

Newly hatched tadpoles have a tail and gills but no legs and no lungs. After a few weeks, the first signs of the complex morphological changes to a frog, called *metamorphosis*, can be seen.

In 1912, the scientist JOHN F. GUDERNATSCH fed thyroid tissue to tadpoles and metamorphosis proceeded faster. If instead the *thyroid* was surgically removed from tadpoles, they grew into giant tadpoles and metamorphosis did not occur. Because of these experiments, scientist examined the thyroid more thoroughly and isolated a substance called *thyroxin*. The addition of only $1/100$ mg thyroxin per litre water is sufficient to start metamorphosis in young tadpoles. Further experiments showed that thyroxin is not effective at all stages of development. Metamorphosis can only be artificially induced by thyroxin when the animals exceed a size of 40 mm. This observation agrees with experimental results showing that the size of the thyroid and, at the same time, the production of thyroxin increases after this developmental stage. From these data, the researchers concluded that a second level of regulation occurs prior to that of the thyroid.

American scientists discovered that tadpoles from which a small gland, called *hypophysis* or *pituitary gland*, on the lower part of the diencephalon (interbrain) had been removed also did not metamorphose. The researchers were able to establish that the hypophysis secretes into the blood a substance that, in turn, stimulates the production of thyroxin. This substance is referred to as *thyroxin-stimulating hormone* (TSH). Despite these successful results, the stimulus of the regulation processes still remained unexplained. Only additional experiments showed that the diencephalon was also involved in the regulation of metamorphosis. Specialized nerve cells produce tiny amounts of a tripeptide and secrete it into the blood stream (*neurosecretion*). This tripeptide is thus called *thyroxin-releasing factor* (TRF). Via the hypophyseal portal system, TRF reaches the anterior pituitary where it stimulates the release of TSH. Only when the hypothalamus has reached a certain size can it react to thyroxin, which then stimulates the release of TRF. This process begins slowly at first and then later becomes a self-enhancing process so that a rapid increase in thyroxin production occurs by the end of metamorphosis (fig. 49.1).

From similar experiments, common principles of <u>hormone</u> regulating can be derived: hormones are produced by cells, tissues or organs and are distributed in tiny amounts in the body. Hormones affect specific target cells, which, in the case of TRF and TSH are located in a respective organ. The target

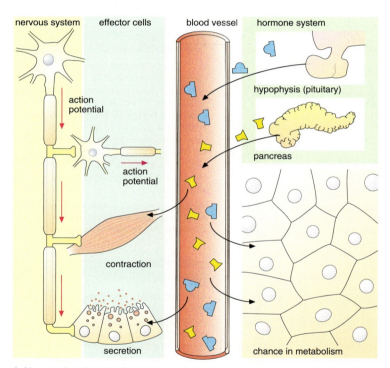

1 Neuronal and endocrine signal transmission

	Hypophysis (pituitary)	Thyroid	Hormone addition	Metamorphosis
Exp. 1	present	present	no	yes
Exp. 2	removed	present	no	no
Exp. 3	present	removed	no	no
Exp. 4	removed	removed	thyroxin	yes
Exp. 5	removed	removed	TSH	no
Exp. 6	removed	present	TSH	yes

2 Release and regulation of frog metamorphosis by hormones cells for thyroxin, however, are distributed throughout the human body. Thus, in addition to the nervous system, the organism possesses a second information and coordination system whose flow of information is slower but is also of duration (fig. 2).

In humans, thyroxin is produced by the thyroid gland located on the sides of the trachea and the oesophagus in front of the thyroid cartilage of the larynx. It causes an increase in basic metabolic rate, especially in the liver and muscle cells. Thyroxin causes a higher enzyme activity in the mitochondria of these cells thereby raising energy conversion. In nerve cells, it activates Na^+/K^+ pumps. If the thyroxin in the blood exceeds a certain level, the secretion in the thyroid is decreased. In humans, thyroxin inhibits TRF and TSH production by a negative _feedback_ system. External factors also influence this regulatory system. For example, a decrease in the surrounding temperature leads to an increase in thyroxin production and consequently the basic metabolic rate rises.

If the function of the thyroid is pathologically increased, metabolism is accelerated too much, which raises body temperature. Even though appetite is boosted and food intake is enhanced, patients lose weight. Heart action is accelerated and restlessness and insomnia occurs. As a therapy, thyroxin is given for a limited time to decrease thyroid function via negative feedback. Since iodised salts are needed for the synthesis of thyroxin, iodine deficiency can lead to a pathologically decreased thyroid function. Basic metabolic rate and body temperature, in this case, fall below the standard values and severe fatigue arises. The patients easily become fat, even though their food intake is small.

Task

① The thyroid of hibernators such as the bat have been shown to become smaller in the autumn. At the end of the winter period, it reached its normal size again. Explain this correlation taking the biological importance of hibernation into account.

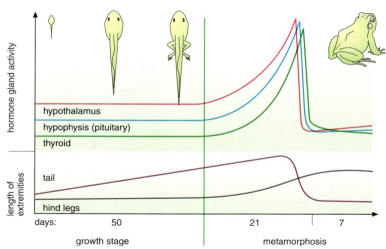

1 Hormonal processes during metamorphosis of the frog

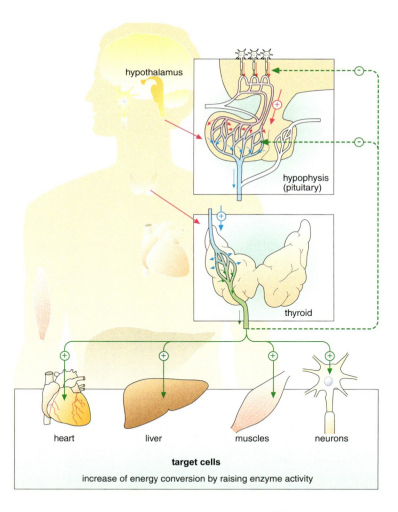

2 Regulation of thyroxin concentration

Neurobiology

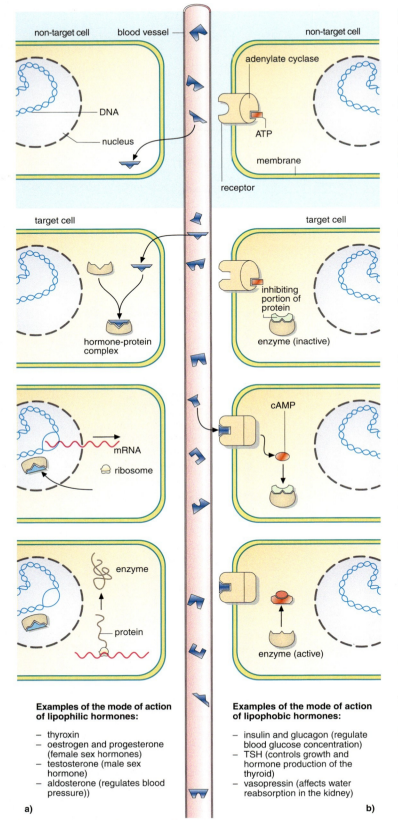

1 Modes of action of hormone effects

Effects of hormones

Hormones differ not only in the nature of their substance but also in their modes of action.

Modes of action

The development of a tadpole into a frog is a genetically regulated complete change of body structure (*metamorphosis*) based on the control of the hormone thyroxin. During this stage, the cells of the tail contain many membrane-enclosed vesicles (*lysosomes*). They store protein-cleaving enzymes (*proteases*) whose concentration is up to 30 times higher than that in the remaining tissue. The proteases degrade the proteins of the tadpole tail. If radioactively labelled thyroxin is added to cell suspensions of various tissues of the tadpole, the radioactivity is detected after a short time in the nuclei of the tail cells. In other cells of the body, it remains in the cytoplasm. Thyroxin is mainly bound to a specific protein in the tail cells. It forms a hormone-protein complex. When analysing the samples after some time, the radioactivity is found to have accumulated in the nuclei of the tail cells.

From this, we can conclude that only this hormone-protein complex can pass through the nuclear envelope and that this complex activates specific genes in the nucleus. The respective mRNAs are produced and translated on ribosomes in the cytoplasm with the resulting synthesis of the associated enzymes, e.g. the protein-degrading enzymes of the lysosomes (fig. 1 a). Thus, the destruction of the tail cell is started. This mode of action is typical for lipophilic (fat-soluble) non-protein hormones that can pass through the biomembrane.

A second group of hormones, in general lipophobic (not fat-soluble) proteins, have a completely different mode of action. These substances cannot pass through the cell membrane. The individual hormone molecules and specific receptor molecules in the cell membrane form a hormone-receptor complex. Here the function of the hormone molecule ends. A single hormone molecule, such as e. g. a glucagon molecule, can cause, within a target cell, a significant change in glucose metabolism within milliseconds. The reaction of a cell to a hormone is called the cellular response (fig. 1 b).

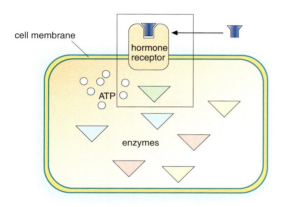

Signal amplification

The extremely high effectiveness of a few hormones belonging to the second type can be explained by an amplification mechanism. A hormone binds to its membrane-bound receptor molecule, and by changing its conformation, stimulates the release of substances called *second messengers* within the target cell. One receptor molecule can trigger the production of several hundred second messenger molecules. These are, in turn, able to start many reactions by activating enzymes. This process does not take place in one step but in 4 sub-steps. During every sub-step, each activated enzyme activates in turn approximately 100 new enzyme molecules. Such an enzyme cascade leads to a 10^8 times amplification (fig. 1). The interaction of a single glucagon molecule with the receptor molecule in the cell membrane finally leads to the production of 100,000,000 glucose molecules from glycogen in the target cell. This cascade reaction is called *biochemical signal amplification*.

The principle of signal amplification via second messengers is not only found with hormones but also in sensory cells and neurons. In synapses, few transmitter molecules can trigger a signal in the transmitting neuron. Even a few fragrance molecules enable us to notice a smell by triggering a cascade in the olfactory sensory cells.

The number of different hormones in an organism is high. In contrast, only a few different second messengers are involved within a cell. Second messengers include cAMP (*cyclic adenosine monophosphate*) or calcium ions. The type of reaction started in the target cell depends on which specific enzymes are activated in that cell. Thus, cAMP can trigger the conversion of glycogen to glucose in muscle cells, the degradation of fats in fat cells or the reabsorption of water in kidney cells. Hormones do not react directly with the adenylate cyclase but with a receptor molecule. Two types of receptor molecules have been found in the cell membrane. Depending on the target cell, they can stimulate or inhibit adenylate cyclase. Hormones can thus indirectly activate or deactivate enzymes in the target cell. This leads to diverse effects in different target cells.

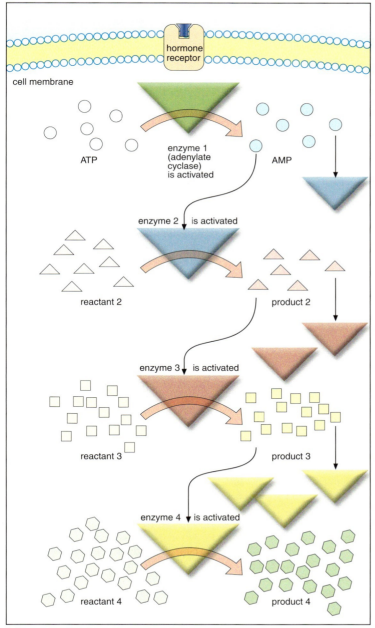

1 Hormone effect occurring via signal amplification by an enzyme reaction

Neurobiology

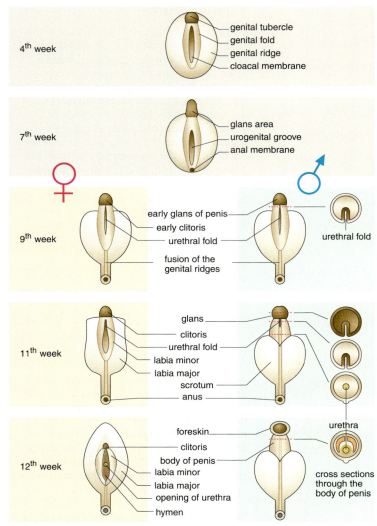

1 External sex organs of human embryos

Hormones and development

In humans, the development of the genital organs can be divided up into two stages. Stage 1: in early embryogenesis, the sex organs can develop into the male and the female form (fig. 1). The Wolffian and Müllerian ducts (primordial gonads) can develop into testicles or ovaries, respectively, and the outer anlage becomes either the clitoris and labia or the testicles and scrotum. The SRY gene, which is usually located on the Y chromosome, determines development into a male. If it is missing from the Y chromosome, an XY woman is born. If the SRY gene is present on the X chromosome, for example because of a crossing over error, an XX man is the result.

Normal development

Stage 2:
Via a gene cascade, the SRY gene leads to the formation of testicles that produce the sex hormone testosterone. During the embryonic phase, testosterone is converted to dihydrotestosterone by an enzyme called 5-α-reductase. It can bind to specific receptors in tissues and thus drive the development of the external sex organs further to form a penis and testicles. The child is therefore born as a boy. If the SRY gene is missing, the development results automatically in a girl. During puberty, testosterone causes a stronger emphasis of the primary and secondary sexual characteristics. However, dihydrotestosterone is only effective prenatally.

Receptors for sex hormones have also been found in many parts of the brain. Here, they cause changes prior to birth and these lead to typical male and female brains.

In consequence, for the normal development of the genital organs, a specific set of genes are important:
— the SRY gene
— a gene, that is responsible for 5-α-reductase, which converts testosterone to dihydrotestosterone
— genes to build testosterone receptors in target cells.

5-α-reductase deficiency

The genes mentioned can, as all genes, mutate, and the loss of function of their gene product is the result. A very rare mutation arises in the 5-α-reductase gene. Since the mutation occurs in 23 related families in the Dominican Republic, it has been possible to research this mutation extensively. In this syndrome, children with an Y chromosome and the SRY gene develop normal testicles that produce testosterone. However, since 5-α-reductase is not functional because of a mutation, the prenatal change of the external sex organs to male genitals does not occur. The sex organs of these children seem mainly female at birth. When, however, a large amount of testosterone is produced during puberty, a change of the sex organs takes place: the clitoris grows into a penis, the labia fuse and form a scrotum into which the testicles migrate in most cases. Unfortunately, since the urethra (which is the tube for both urine and sperm) ends at the root of the penis in

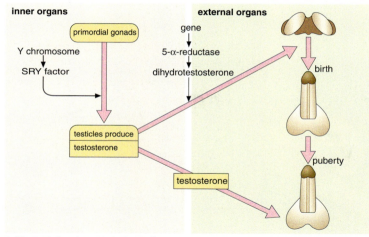

1 Normal development of a male

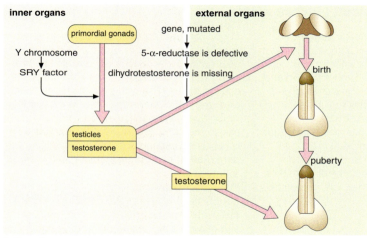

2 Sex change

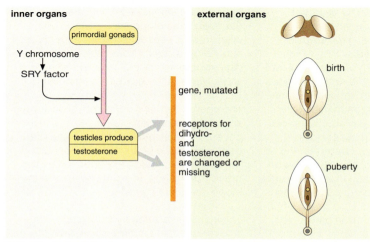

3 Testicular feminization

these men, they are not able to reproduce. The indigenous people call these children "machihembra" (man and woman in one).

In 19 of 33 examined cases, the affected children were clearly raised as girls. After the changes of their sex organs, all except for one changed sexual identity; thus, they thought of themselves as male and were interested in women as partners. Today, most of them live in traditional marriages. These men can therefore fulfil the male gender role that is expected of them by society. We can see that, at least in the tolerant conditions of the Dominican Republic, the influence of the sex hormone (testosterone) has a greater effect on male gender identity than 12 years of education as a girl.

Testicular feminization

Another problem during development arises if a mutation in the gene for the testosterone receptor causes the receptors to be missing in the target tissues or to be non-functional. In this case, the SRY factor causes the formation of the testicles, which begin to produce normal amounts of testosterone but, since the necessary receptors in the target tissues do not function, the change of the external sex organs cannot start and the development of the male characteristics during puberty is absent. The affected people remain phenotypically women. People with this mutation cannot be distinguished by their looks from a normal woman. They have a blind ending vagina and have testicles as inner sex organs. This anomaly is often only found by a doctor when the patient visits because of absent periods.

Tasks

① Define the terms "genetic gender", "phenotypic gender", "gender identity" and "gender role".
② What brain development is expected in girls with a sex change?

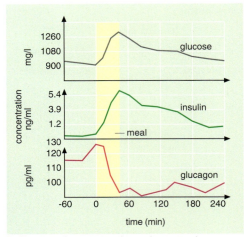

1 Testing blood sugar concentration

2 Change of concentrations after a meal

Regulation of the blood glucose level

Nerve cells cover their energy demand almost entirely by glucose. In humans, a limited supply of glucose can lead to a decrease in the power of concentration and, in extreme cases, to a loss of consciousness. Nerve cells can store only a small amount of glucose. Blood continuously delivers new supplies of glucose. The sugar content in the blood of an adult is about 5.4 g. A rough estimation shows that this amount is enough to sit calmly for 12 min or to ride a bike for 2 min. After this, we would have to eat.

Even though food intake occurs periodically and energy demands vary, depending on physical activity, the blood glucose concentration in a healthy person fluctuates only little. Thus, we can conclude that, in our body, we have a regulatory system that keeps the blood sugar concentration constant. This is confirmed by a so-called *oral glucose tolerance test* during which a healthy person drinks 75 g of glucose dissolved in water. The sugar concentration in the blood is then measured in defined intervals. The blood sugar concentration rises for a short time up to a maximum of 2000 mg per litre blood, sinks within 2 hours and levels off again at the original concentration (fig. 2). 75,000 mg glucose in about 6 litres of blood would be expected to result in a blood sugar concentration of 12,500 mg per litre blood.

During food intake, and thus in parallel to the increasing blood sugar level, a hormone is released that causes a decrease in the sugar content (fig. 2). This hormone, which is called *insulin*, is produced in the pancreas, in island-like groups of cells that can be easily distinguished from the rest of the pancreatic tissue. The groups of cells are called, after their discoverer, the *islets of Langerhans*. Every increase in blood glucose level above the set value (about 900 mg/l blood) is recorded in special cells, the β-cells, of the pancreas. They secrete insulin into the blood.

Insulin has a diverse mode of action. It supports glucose uptake into cells, especially into muscle and fat cells. In the liver and muscle cells, it stimulates glycogen synthesis (glycogenesis) and energy conversion by glucose degradation. At the same time, it inhibits glycogen degradation. The syntheses of fats and proteins from glucose are stimulated. Thus, insulin decreases the blood sugar level in two ways: first, it causes the storage of glucose and, second, it stimulates its consumption (transport through the cell membrane and an increase of the glucose metabolism).

The energy demand during physical activity and even the basal metabolic rate is mainly covered by energy derived from glucose degradation. If a lack of glucose occurs in the cells, it is immediately supplied by the blood. As a result, the glucose level in the blood decreases.

The insulin secretion by the β-cells is then also reduced and, after a short time, the insulin concentration decreases because

Blood sugar level
A blood sugar level of 90 equals 90 mg/dl = 900 mg/l blood.

Rough estimation:
About 6 litre of blood of an adult contain
6 · 900 mg = 5.4 g glucose.
180 g glucose = 1 mol ≙ 2836 kJ
5.4 g glucose = 85 kJ
energy demand during cycling: 40 kJ/min.
The energy content of the blood sugar is sufficient for about 2 min.

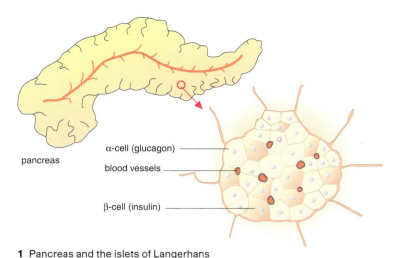

1 Pancreas and the islets of Langerhans

insulin is rapidly degraded in the body. Thus, its inhibitory effect on the release of a second hormone, *glucagon*, which is produced in the α-cells of the islets of Langerhans (fig. 1), is reduced. Glucagon stimulates the degradation of glycogen in the liver and, as a result, the synthesis of glucose. Proteins and fats are also degraded for glucose to be synthesized. These processes lead to a rise of glucose concentration in the blood. Insulin and glucagon are often described as antagonists because they inhibit the release of each other and stimulate opposite processes. However, when examining this more closely, we can see that they complement each other: glucagon is responsible for the release of stored glucose from the liver, and insulin for the decrease of the glucose concentration in the blood.

If the parameters in a system are set in such a way that they remain at about the same value, this is referred to as *regulation* (fig. 2). The regulated parameter is the *control variable*, in this case, the blood glucose concentration. A change in blood glucose concentration by physical activity or food intake is referred to as the *interference variable*. Pancreatic β-cells contain measuring points, the so-called *sensors*, that make the precise measurement of the blood glucose concentration possible. The measured value is referred to as *actual value*. The set and actual values are compared by a controller, in this case, the diencephalon and the α- or β-cells. If the two values differ, the blood glucose concentration is regulated by *effectors*. These effectors are the muscle or liver cells that take up the glucose or release it. The release or uptake is represented in the regulation schema by the *correcting variable*.

In extremely demanding situations such as fight and flight and also in stressful situations, a third hormone joins in. This is *adrenalin* (*epinephrine* in US), which is produced by the adrenal medulla. In contrast to glucagon, adrenalin stimulates glycogen degradation in the liver and muscles via the sympathetic nervous system. In this way, it causes fast energy release from storage depots in the body.

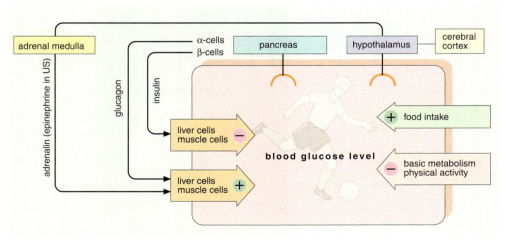

2 Simplified regulation schema

Issues

Diabetes mellitus

The English physician THOMAS WILLIS (1621 — 1675) was the first to report the honey-like taste of the urine of diabetics. From this, the name of the disease was derived: *diabetes* means "to pass through", *mellitus* "honey-sweet". In 1835, the Italian physician AMBROSIANI isolated sugar crystals from the urine and from the blood of diabetics. At that time, the chemical structure of glucose was unknown. Sugars were "sweet-tasting substances".

Pioneers in the research of blood sugar regulation were J. VON MERING, O. MINKOWSKI, F. GRANT BANTING und C. BEST who carried out experiments with the pancreas from dogs. They showed that this organ is essential for the regulation of the blood sugar level.

Find out more about the results of these researchers. Draw a timeline in the form of a poster to demonstrate the development of their ideas.

Sweet proteins

Thaumatococcus daniellii is a plant in the tropical rain forest of West Africa. Its fruits are used to sweeten bread, tea and palm wine. The sweet taste is, however, not caused by sugar but by a protein called *thaumatin*. By using bacteria, the gene responsible for thaumatin synthesis has been transferred to potatoes. This makes it possible to use the protein in large-scale industrial applications. Attempts have also been made to transfer this gene to fruit-bearing plants such as strawberries, melons and apples.

Teeth also benefit! Explain the advantages of this sweetener in contrast to common products.

More information regarding the terms thaumatin and thaumatococcus can be found online. Collect the most important factors.

Bread/potato group
1 BE relates to:
15 g zwieback
30 g bread rolls
80 g potatoes
100 g peanuts
220 g yoghurt
100 g apple with peel
160 g blueberries
250 g water melon

Explain the term bread exchange (BE) and its importance for diabetics. BE lists are available in pharmacies, at the doctor's or online. Calculate the ingested BE units for one day.

Karlsbad was an important therapeutic bath for diabetics in the 19th century

Glucose value

The glucose value in the blood has to be correct. Thus, a diabetic has to be able to measure easily and rapidly his/her blood sugar levels. Modern measuring devices have a set of microcannula (microneedles) that are filled with blood. For this, only 3 µl of blood is required.

Collect facts about up-to-date measuring methods. Information for this topic can be found at the doctors', in pharmacies or online. Continuous glucose monitoring is the future aim of diabetes research. It should become possible to regulate normal blood glucose levels automatically via an insulin pump that has a glucose sensor. State the advantages and explain these by using a regulation loop (see p. 55).

Blood glucose level

Sports, food, work and insulin influence the blood glucose concentration. In addition, stress and other emotions also play a role. Diabetics often suffer from being dependent on insulin and from constantly having to calculate their energy values. The body of a sufferer does not automatically regulate these values so that a diabetic has to adjust them himself all the time according to the demand. He has to learn to distinguish between reliable and non-reliable symptoms in order to inject the correct amount of insulin at the correct time. If a diabetic does not pay attention to the symptoms or if he measures his blood sugar incorrectly, hypoglycaemia or hyperglycaemia can occur.

Work out the biological correlations. A report about an affected person is helpful for this project.

For centuries the inhabitants of the Pacific island Nauru were able to obtain food only laboriously by fishing or simple agriculture. However, since massive phosphate resources were discovered on the island, the island's inhabitants have become wealthy. Unfortunately, the jump from an existence as a hunter/gatherer to modern civilisation has had its price. The lack of movement and food in excess has led to obesity in the majority of the population. The number of diabetics has increased dramatically. Today, every third inhabitant has diabetes.

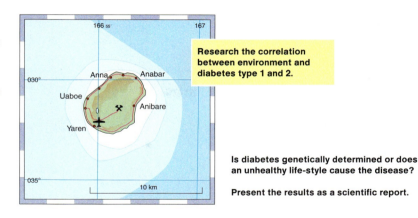

Research the correlation between environment and diabetes type 1 and 2.

Is diabetes genetically determined or does an unhealthy life-style cause the disease?

Present the results as a scientific report.

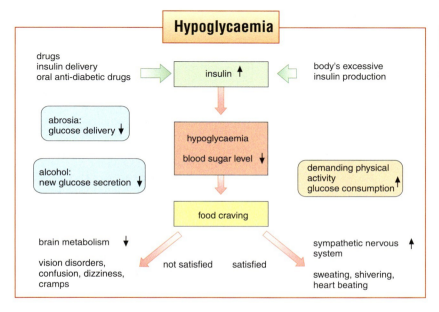

Blood glucose level

If the blood glucose level sinks below 0.45 g/l (*hypoglycaemia*), the functions of the brain are disturbed. The person becomes helpless and unconscious and experiences seizures. Injury often occurs because of falling or traffic accidents. It is thus important for **type 1** diabetics to recognize hypoglycaemia in time.

Demonstrate the difference between hypoglycaemia and hyperglycaemia.

What are the reasons for subsequent injury?

Explain the different disease symptoms and show how diabetics can protect themselves.

People are having fun at a birthday party. Everyone is dancing and plenty of alcoholic drinks are available. All of a sudden, Martina feels very weak. Her friends do not know what to do. She did not seem to be drinking that much. Then they remember that Martina is diabetic. How can they help her? It is difficult to distinguish hyperglycaemia from a hypoglycaemia.

Gather information about the signs of hyper- and hypoglycaemia. Find out how you can help in such a situation.

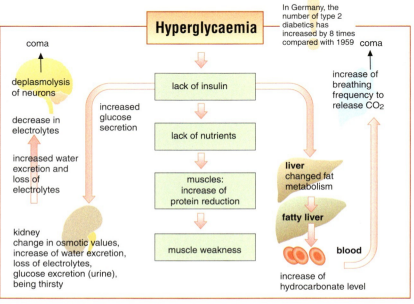

Neurobiology

Stress

Tupaias (tree shrews) are small mammals in the rainforest of South East Asian and are active by day. They live alone or in pairs in territories that they defend fiercely against intruders. The secretion of a gland (neck gland) is used by sexually mature males to mark their territory and by females to mark their offspring.

Usually, the tail hair of the tupaia lies flat. However, if two sexually mature males meet, this hair is raised and they fight each other. If the animals are separated after a fight by using a wooden wall, the loser and the winner recover within the same time. DIETRICH VON HOLST separated two fighters by only using a wire mesh. The heart rates of both animals were recorded with an implanted mini-transmitter. In parallel, VON HOLST determined the percentage of the time during which the tail hair remained raised. In the winner, both parameters sank to normal values shortly after the fight. In the loser, the sight and probably the smell of the winner led to a constantly higher heart rate and almost always to raised tail hair. The length of the duration of the raised tail hair is a visual parameter of the internal agitation of the animal (fig. 1).

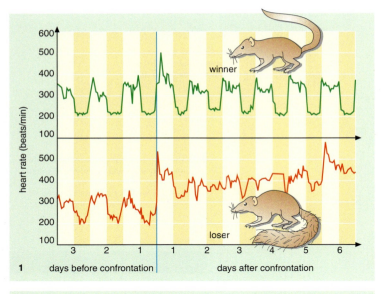

1 days before confrontation / days after confrontation

By raising their tail hair, tree shrews not only react to foreign males in the same cage, but also to an increased population density (*social stress*), to suddenly occurring noises or the sight of a strange individual, i.e whenever something unknown and frightening approaches the animal. The animal is said to be in a *stress situation*. DIETRICH VON HOLST observed that, if such situations continue for a long time, they cause psychological, physical and ethological changes (fig. 2). If the animal cannot avoid the stress-causing factors, weight loss and death are the results.

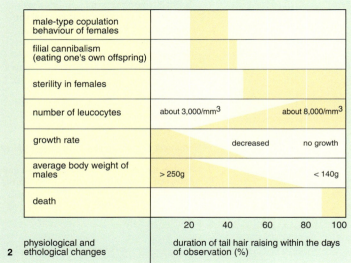

2 physiological and ethological changes / duration of tail hair raising within the days of observation (%)

If adrenalin (epinephrine in US) is injected into a tupaia, it raises its tail hair. This and similar experiments have revealed the physiological processes occurring within the animal. Optic, acoustic and olfactory stimuli are received and processed in the cerebrum (fig. 59.1). From the cerebrum, nervous impulses reach the *hypothalamus*. The hypothalamus in turn stimulates the sympathetic nervous system as part of the autonomous nervous system. The sympathetic nervous system transmits impulses to the heart, blood vessels and adrenal

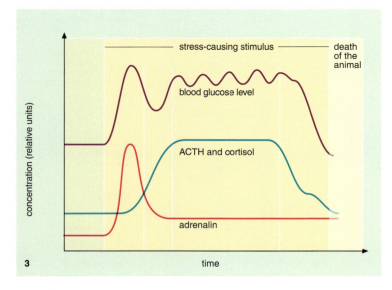

3 time

58 Neurobiology

burdens and is placed into a state of stress. This is the original meaning of "stress". Thus, stress is an adaptation and consequently a vital reaction of the organism to external stress factors, called *stressors*.

Stressors act via the hypothalamus, which secretes more releasing factors. These stimulate the pituitary gland to secrete more ACTH (*adrenocorticotropic hormone*), which itself acts on the adrenal cortex. Animals in a state of stress have an enlarged adrenal medulla, which releases more hormones, including *glucocorticoids* such as cortisol. These increase the resistance power of the body by, for example, inhibiting protein synthesis and stimulating protein degradation in the muscles, bones and lymphatic tissue. In this way, more free amino acids reach the blood stream and are used to produce new glucose in the liver. This increases the blood glucose level. By inhibiting protein synthesis in the lymphatic organs, glucocorticoids have an anti-inflammatory effect because reduced antibody production leads to a slower defence response against infections. All these changes, which only occur after a certain time, are called *general adaptation syndrome* (*GAS*).

Occasional stress with appropriate recovery periods can increase the resistance power of the body and is called *eustress* (fig. 1). If recovery periods are absent, the occurring continuous stress is called *distress*, which can be dangerous for the organism. A continuously elevated adrenalin level resulting from frequently occurring fight-and-flight syndrome, however, also causes increased ACTH secretion and thus stress. This is why today we refer to agitation, even if it only occurs once, as "*stress*". Continuous such pressure on the entire organism can lead to cardiovascular diseases after several years. Heart infarction is one of the most common causes of death today.

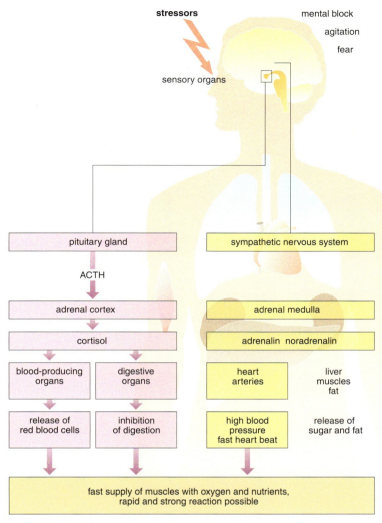

1 Cooperation of nervous and hormone system during eustress

Task

(1) State the similarities and differences between fight-or-flight syndrome and stress.

Glucocorticoids
Non-protein hormones that can dissolve in fat (lipophilic), e.g. cortisol.

medulla. The heart rate rises and blood pressure increases. In the adrenal medulla, more adrenalin is released. Narrowing of the blood vessels, except in the skeletal muscle, is one of the various effects of adrenalin. Thus, blood pressure rises. Sugar is mobilized via glycogen degradation in liver and muscles and hence the blood glucose level rises. The biological importance of these reactions is the fast activation of energy resources in an attack or for flight, so that high performance is possible. These adaptations of the body are called *fight-or-flight syndrome*.

The body can also adapt to harmful influences such as heat, cold, hunger, thirst, infections, wounds or psychological

1 Grapes with/without gibberellin treatment

2 Plant with/without gibberellin addition

Plant hormones — phytohormones

Agar
A gelatine-like substance that is made without boiling.

Phytohormones
= plant hormones

In the control of their development, the germination, growth, flowering phase or falling of leaves plants depend on external factors such as light and temperature. Chemical substances distribute the effect of the external factors within the entire plant. These substances are the plant hormones). In contrast to the hormones of animals and humans, *phytohormones* are not produced in a specific organ but in the cells of various tissues. Their function is also not strictly specific; a phytohormone can affect diverse processes and inhibit or stimulate them (fig. 3). The cascade-like effect in the cell can be started even by small concentrations of the hormone and is identical to that in animals and humans. Phytohormones affect gene regulation in the respective plant cells and stimulate or inhibit the production of certain enzymes (see page 50 / 51).

Phytohormones control growth

The first ideas on the regulation of growth by phytohormones were based on the experiments of CHARLES DARWIN. In his work "The Power of Movements of Plants" in 1881, he examined the process of *phototropism* on oat seedlings. Hereby, he tried to answer two questions:
a) which part of the plant reacts to light and
b) which factor makes the plant direct its growth towards the light?

From his experiments, in which he removed the tip of the seedlings or covered different parts of the seedling with light-proof material, he came to the conclusion that chemical substances are produced in the tip of the seedling and are then distributed to the remaining tissue (fig. 61.1 a).

In 1913, PETER BOYSEN-JENSEN extended the investigation of these questions. He removed the tip of the oat seedling and observed that the remaining tissue no longer reacted to light. Based on this observation, he placed a piece of Agar on top of the cut surface; this did not cause any changes. After first placing the tip of the seedling onto the piece of Agar and then placing the Agar on the oat seedling, the seedling started to react to light once again (fig. 61.1b). Further experiments were performed in 1913 by ARPAD PAAL. He removed the tip of the seedling and replaced it but shifted to the side. Growth occurred only on the side on which the tip was attached. This led to a bending to the

Hormone	Effect
Abscisic acid	inhibits germination, closes stomata, promotes the aging of leaves
Auxins	promote length growth and cell division, inhibit flower initiation and falling of leaves
Cyto-kinins	promote bud formation and leaf unfolding, open stomata, delay the aging of leaves
Ethylene	promotes fruit-ripening and aging of leaves, inhibits length growth
Gibberel-lins	promote cell division in the shoot, promote leaf unfolding and tuber production (potato), promote germination

3 Phytohormones and their effects

side where no chemical substances were released into the tissue. PAAL concluded that bending in phototropism is attributable to different concentrations of the chemical substance and that, depending on the concentration, this leads to stronger or weaker growth (fig. 1c).

Additional experiments by FRITS WENT in 1926 established the existence of a substance as carrier of information. WENT removed the tips of the oat seedlings and placed them on pieces of Agar. He suspected that the chemical substance or substances would diffuse into the Agar pieces. If now these Agar pieces were placed on the seedling without the tip but shifted to one side, a clear bending was the result (fig. 1d). He called the suspected substance *auxin*, which is the Greek word for "growing". In 1933, the substance was isolated and its chemical structure examined. It was identified to be an *indole-3-acetic acid* (*IAA*; see margin). The chemically produced substance has the same effect as auxin and is used for plant breeding. New investigations have shown that IAA affects fast gene regulation and, for example, causes the synthesis of additional cell wall material.

Phytohormones control fruit ripening

Phytohormones are not only transported dissolved in the tissue fluid, but also as gas. A gaseous phytohormone is *ethylene* (ethene). Ethylene promotes the ripening of fruits and has many other effects (fig. 60.3).

Fruits such as bananas are gathered unripe because of their long transport routes. After transport by ship, they are kept in German storage halls in the presence of ethylene. This accelerates the ripening process. However, the disadvantage is that ethylene pushes ripe fruits to an even riper state and then into faster decay. In gene-modified tomatoes that do not become soft so fast, ethylene synthesis has been reduced by gene modification. However, ethylene is not only important for fruit ripening, but also during germination and growth processes. Here, it inhibits length growth.

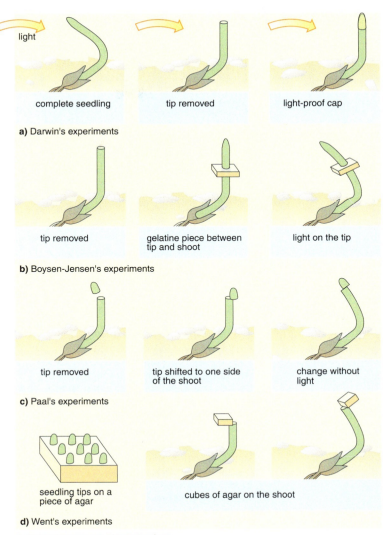

1 Experiments on phototropism

Tasks

① Use figure 1 and the text to summarize the scientific findings of each experiment and state which questions regarding phototropism and the importance of phytohormones remain unanswered.

② Gather information about the importance of ethylene for the flower industry and make a short report.

Structural formula of indole-3-acetic acid (IAA)

Neurobiology **61**

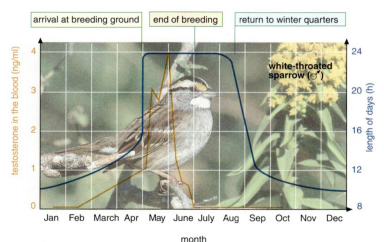

1 Breeding behaviour and hormones of the white-throated sparrow

Hormones and behaviour

The drive to migrate and to carry out brood-care behaviour in the white-throated sparrow that breeds in Alaska and overwinters in California has been shown to be influenced by hormones (fig. 1). In contrast to the time-limited local effect of the nervous system, hormones are distributed throughout the organism. This means that they have a systemic and thus longer-lasting effect. In general, a change in hormone level only influences the readiness to act. It does not influence the form or the way that the behaviour is performed, but only the frequency and intensity of the occurrence in situations triggered by a specific stimulus.

That the frequency of a certain behaviour really depends on a hormone can be examined in experiments by, for example, measuring the hormone level in the blood. A hormone gland can also be surgically removed and the effects on the behaviour can then be examined (fig. 2). When the hormone is artificially given later, we can establish whether the behaviour can still be carried out if hormone is present.

Hormones often have many different effects. *Oxytocin*, a hormone present in all mammals, is released in sheep during birth. It has a labour-inducing effect and additionally stimulates the bonding between mother sheep and newborn. If the lamb is separated from the mother directly after birth, it is usually not accepted by her later. After the artificial administration of oxytocin, care occurs once again. In the Prairie Vole, oxytocin supports bonding within a mating pair and decreases the frequency of aggressive behaviour.

In various regions of the mammalian brain, receptors have been found that lead to a higher excitability of some neurons in the presence of oxytocin, whereas in others, excitability is decreased. Oxytocin receptors often occur in different densities even within the brains of animals of the same species. An accordingly variable responsiveness to the hormone has been observed experimentally. Hormones greatly influence behaviour, although other factors are also important. For example, the stimulating effect of oxytocin on brood-care behaviour only occurs during mating season.

Task

① During courtship, the zebra finch male constantly sings a short courtship song. Its frequency is a good measure of its readiness for courtship. The testicles were removed from a group of animals. The control group underwent a sham operation (i. e. all surgical procedures were carried out but the testicles were not actually removed) The courtship behaviour of both groups was then compared.
   a) Describe and interpret the results shown in figure 2a.
   b) The castrated animals were given testosterone 4 times at short intervals. Interpret the results (fig. 2b)?

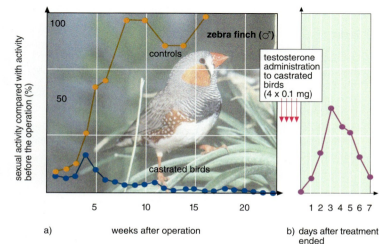

2 Castration experiments in zebra finches

Neurobiology

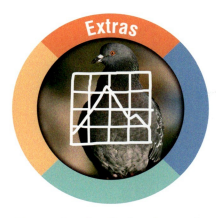

Reproductive behaviour of the Barbary Dove

The American behavioural scientist DANIEL S. LEHRMANN examined the reproductive behaviour of the Barbary Dove, including the influence of hormones.

1. Set of experiments

1. LEHRMANN placed a pair of Barbary Dove males and females that had previously been kept separate into one cage each and, 7 days later, offered them material to make a nest and a finished nest with two eggs in it. None of the animals started making a nest or brooding (i. e. sitting on the eggs to make them hatch).
2. One male and one female that had previously been kept separate were placed together in a cage that contained a finished nest with two eggs in it (fig. 1a).
3. In order to find out whether the birds first had to get used to the cage, he placed the birds in pairs into the cage for 7 days but separated them with a panel of frosted glass, which was removed on the 7th day (fig. 1b).
4. In a further experiment, LEHRMANN put a pair of doves into an empty cage and, 7 days later, gave them a nest with two eggs (fig. 1c).

Task

① Conclude from the experimental results which stimuli trigger nest-building and brooding and explain your statements.

2. Set of experiments

The difference in time between offering the stimuli and subsequent action leads

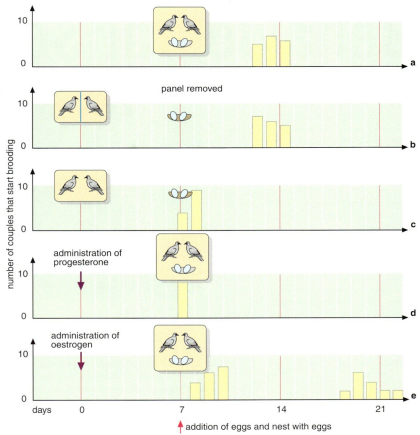

1 Reproductive behaviour of Barbary Doves

to the hypothesis that the stimuli do not directly trigger the action but stimulate hormone glands.
In order to test this hypothesis, LEHRMANN gave dove females progesterone or oestrogen. The conditions of the experiments and the results are shown in figure 1d and e.

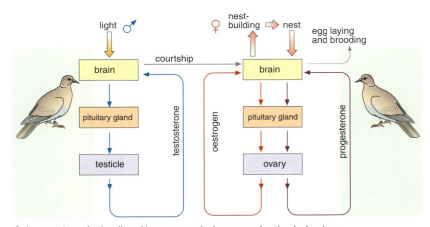

2 Interaction of stimuli and hormones during reproductive behaviour

Tasks

② Interpret the experimental results (fig. 1d and e).
③ What is your opinion of LEHRMANN's hypothesis regarding the hormone effect on behaviour?
④ Explain all experimental results presented in figure 2.

Neurobiology **63**

Behavioural Science

Behavioural science includes all human and animal behavioural patterns that are scientifically examined by using biological methods. "Behaviour" is thus all the observable details of the movements, body posture and sounds of an animal, including when it is at rest. Reactions to stimuli are distinguished from spontaneous actions and often the questions arise regarding if and to what extent behaviour is changed by experience.

The question as to why an organism in a specific situation acts according to a specific pattern can be related to various categories. In the early days of behavioural science, congenital aspects (inborn) were distinguished from learnt aspects. Animal behaviour was strictly separated from that of humans and everything was analysed with regard to the principle of species survival.

Today, the question is allocated to two greater complexes. On the one hand, direct connections (_proximate causation_) are sought. Physiological mechanisms that depend on internal conditions and trigger stimuli are examined, as are processes that originate in development. One possible question is: how can certain neuronal networks or hormone effects cause behaviour?

Information is derived from observations in the wild and from laboratory experiments.

The initial question is: What kind of behaviour is occurring here?

On the other hand, basic connections (*ultimate causation*) are sought. Here, phylogenetic development and ecological adaptation or adaptation within one species are the main focus. They are often examined by using a cost-benefit analysis. Why have certain neuronal networks and hormone effects been successful in evolution? What advantages do they offer to the individual?

> How does behaviour act at a specific moment and what stimulates this behaviour?

Such phenomena are examined in *sociobiology*. This branch of behavioural science has shown that a specific type of social behaviour depends on the ecological conditions that were originally present when it developed during evolution. New questions regarding the reproductive fitness of an individual or the importance of altruism (apparently selfless acting) also change the way that we look at human (and animal) behaviour today.

> Why does a specific behavioural pattern work "in this way and no other" and what does it achieve?

65

1 Issues in behavioural science

Causal and functional problems

Animal behaviour is examined for various reasons, e.g. to save a species threatened by extinction or to find optimal conditions for a farm animal. In general, behavioural scientists are interested in finding basic principles that can be applied to behavioural patterns of many animals and even of humans. These research focuses have different starting points.

Monarch butterflies

The example of the monarch butterfly

These large butterflies, which are almost 10 cm in size, have developed in close association with poisonous plants (milkweed family) that exist from the tropics to Canada. After mating, the female lays hundreds of eggs on the food plant. The strikingly coloured caterpillars shed their skins five times within 14 days. The toxic substances that are contained in the leaves of the milkweed do not harm the caterpillar; instead, they are stored in their cells as the result of a complicated metabolic pathway. The animals hence themselves become toxic, which will later protect the adult butterflies from predators. Freshly hatched butterflies are able to reproduce straight away and thus a second and third summer generation is generated. The last generation hatches under conditions of shorter days and colder nights. These animals do not become immediately sexually mature. They have stronger flying muscles and soon start on their migration to their winter quarters.

In North America, three enormous populations are distinguished (fig. 67.1). The West Coast butterflies from the mountains of California, Oregon and Washington migrate to the cool and moist coast of California. They hibernate in the few eucalyptus and coniferous forests. In spring, the butterflies wake up and exhibit a genetically fixed mating ritual. They then spread northeast to the mountains.

In swarms of several trillion animals, the second population migrates from central North America to central Mexico in the late summer. In the Sierra Madre, they fly up into valleys at an altitude of 3,000 m with high air humidity and low temperatures. Here, they overwinter on Mexican fir trees in about 30 places of only a few hectares in size. The middle parts of these trees are almost completely covered by butterflies, some even sitting on top of each other. They remain for up to 5 months in such "sleeping communities". In the middle of March, the now sexually mature butterflies fly down into the valleys like slowly flowing streams. Here, they drink nectar from the spring flowers and take up urgently needed water before flying back to the central regions of North America. Mating occurs in the winter quarters and during the return flight. Egg disposition can happen even during the return journey so that the migration to the north is also continued by the newly hatched offspring.

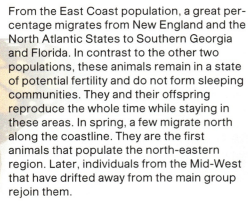

From the East Coast population, a great percentage migrates from New England and the North Atlantic States to Southern Georgia and Florida. In contrast to the other two populations, these animals remain in a state of potential fertility and do not form sleeping communities. They and their offspring reproduce the whole time while staying in these areas. In spring, a few migrate north along the coastline. They are the first animals that populate the north-eastern region. Later, individuals from the Mid-West that have drifted away from the main group rejoin them.

1 Monarch butterfly

These three monarch populations can reproduce among each other and have fertile offspring. Thus, they belong to the same species according to the genetic definition of a species.

Proximate and ultimate causations

The behaviour of the butterflies can be examined by using various biological disciplines. One starting point for behavioural research can be, for example, the fact that the butterflies migrate over long distances of up to 5,000 km and still find their exact places to overwinter in Mexico.

Neurophysiologists and behavioural scientists examine how the animals can achieve this and what stimuli or genetic factors trigger or influence this behaviour. Using their inner clock, the butterflies, like bees, follow a sun compass. These findings of the *causative connections* that trigger and control a specific behaviour in an animal are collectively called *proximate causations*. Such studies are grouped under the heading of *Ethology*.

Other behavioural scientists and ecologists question the advantages of migration behaviour against staying in the same place. Many non-migratory butterfly species overwinter as larvae or pupae in a protected place and thus survive poor climate conditions and reduced food supplies without the dangers and energy expense of long migrations. The kind of adaptation that is the base for migration behaviour is studied under the heading of *Behavioural Ecology*.

The summer generation and the migrating butterflies have different mating behaviour and a different social organization. In summer, each animal is sexually active immediately after leaving the pupa. Straight after hatching, the females receive, from the male, a large spermatophore that not only contains sperm, but also nutrients for the female. This process takes almost 6 hours and is highly complex. A female that mates three to four times thus obtains a significant percentage of nutrients that are essential for laying eggs and for survival. Independent from the triggering mechanism for this behaviour, one can ask what advantage this behavioural strategy has over others. If we assume that, during evolution, those animals who had more offspring exhibited better adapted behaviour, then we can ask: what is the function of this behaviour and

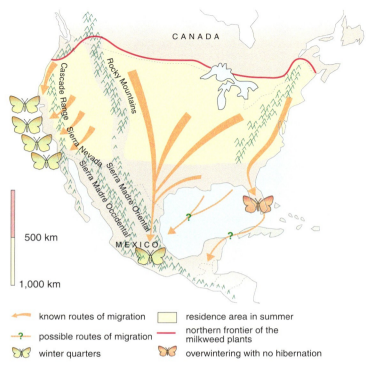

1 Migration routes of the monarch

what is the reproductive advantage for the respective animals? These *ultimate causations* of behaviour are examined under the heading of *sociobiology*.

Tasks

① Explain broadly the terms proximate and ultimate causations of behaviour.
② Analyse the example of the monarch butterfly and list possible proximate and ultimate causations of individual patterns of behaviour.
③ Different indigenous animal species such as the hedgehog and squirrel also show seasonally dependent changes in behaviour. What similar proximate and ultimate problems do they need to solve?

The history of behavioural science

Three great eras characterize the history of behavioural science: The early ancestors of today's humans observed the behaviour of animals and adapted their hunting strategies based on traditional knowledge and tamed animals as companions or helpers. In the subsequent pre-scientific phase, simple observations were mixed with religiously influenced interpretations. Animal behaviour also formed the basis of fables. Observers of nature, who precisely described the behaviour of individual animals, only questioned what they had observed in the 18th and 19th centuries. This led to the era of scientific behaviour analyses containing several areas of focus.

Classical ethology

In Europe, animal psychology developed from classical ethology mainly founded by OSKAR HEINROTH and KONRAD LORENZ. It was originally based on the concept of evolution by CHARLES DARWIN and developed into the *theory of instinctive behaviour*. The supporters of this theory believed that many behavioural patterns are congenital (present at birth) and thus are carried out in the same way in all individuals of one species. This hereditary coordination was thought to be a part of instinctive behaviour and allowed comparative behavioural research. Animals with behaviour differing from the standard were considered as not relevant by LORENZ (type theory). Instinctive behaviour, according to this theory, is carried out by animals that are ready to act in a specific manner

appetency

leaving its hiding place, waiting state

taxis (tactic movement)

directed turning towards or sneaking up to an object

consummatory act

fixation with both eyes and snapping

further reactions

swallowing, mouth wiping

Prey catching behaviour of the common toad

and is triggered by specific key stimuli. The requirement for the filtering of the excitation pattern should be a congenital *fixed action pattern* (FAP). Examples are the prey catching of the common toad or the egg roll movement of the greylag goose.

In 1951, NIKOLAAS TINBERGEN wrote the book "The study of instinct". In this book, a hierarchy of congenital fixed action patterns and of motor centres was claimed to be associated with the reproductive behaviour of the three-spined stickleback. Attempts were made to combine observations of behaviour with results of neurophysiological examinations. Another approach was chosen by the Austrian zoologist KARL VON FRISCH. By using conditioning experiments, he could establish that bees have a sense of colour and the ability to see ultraviolet light. Furthermore, the decoding of the bee language showed how complicated an inherited social communication system could be. In 1973, KONRAD LORENZ together with KARL VON FRISCH and NIKOLAAS TINBERGEN received the Nobel Prize.

Behaviourism

In 1912, the American biologists J. B. WATSON and E. L. THORNDIKE founded a type of behavioural science that avoided subjective terms and drew no conclusions about the inner state of the animal. This so-called *behaviourism* contributed greatly to the theory of learning (see page 80).

1 Konrad Lorenz and his geese

2 Greylag goose

The instinct doctrine — coming under criticism

For 50 years, KONRAD LORENZ and the instinct doctrine have lead to widespread controversy and wide-ranging research.

The term instinct

The term "instinct" has an emotional aspect and is used diversely and imprecisely. The "instinctively correct reaction" of the car driver in a dangerous situation and the statement that animals react "automatically correctly" are both examples of unsupported reasoning. Early in ethology, the term *instinctive behaviour* was used instead of *instinct*. It described a genetically determined behavioural pattern with an orientation and a motion element. The orientation element (taxis) during prey catching is the guided movement towards the prey together with the movement element (*consummatory act*) involving fixation and snapping. Before this takes place, an undirected search (*appetence behaviour*) can occur. This increases the probability of finding prey during the search for food.

Since in only a few behavioural patterns following a strict scheme was involved, an *instinct conditioned interconnection* was assumed for processes such as obligate learning (example: an inexperienced squirrel first nibbles randomly and then improves its technique to get at the food inside a nut). Today the term "instinct" is not usually used since congenital (inborn) and learnt behaviour do not necessarily represent a contradiction.

Readiness for action

LORENZ invented the term *action-specific energy* to describe motivation. This challenged the critics, especially with regard to the psychohydraulic model. In this model, an excitation increases from internal and external factors until the threshold of the respective congenital activator mechanism is passed or until a key stimulus occurs that triggers the consummatory act or *final action*. The quality of the activating stimulus and the inner condition determines the intensity of the action (*law of dual quantification*).

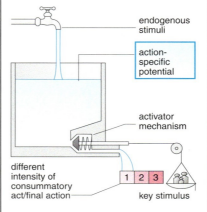

Many behavioural and neurophysiological experiments could not show an instinct centre in the brain. The probability that a certain behaviour is executed depends, among other things, on the time of the day or year, the endogenous rhythm, the hormonal state, the availability of supplies, the state of health and the age of the animal, ecological conditions and, mainly, the animal's previous experience.

The key stimulus

This term is derived from the concept that a congenital *fixed action pattern* (FAP) filters incoming stimuli and precisely recognizes the pattern that fits. Social key stimuli that serve an unambiguous communication were termed as activators and preferably analysed by decoy experiments (e.g. the red belly of a stickleback). With examination methods becoming more exact, researchers could show that, on the one hand, not a single stimulus but rather a complex pattern of stimuli and stimuli constellations trigger actions. On the other hand, although the neuronal mechanisms of information processing are not completely decoded, they are more flexible than a simple lock-and-key relationship.

Vacuum activity

If an animal has not performed an activity for a long time, an accumulation of "readiness of action" makes it possible that weaker stimuli can trigger the action. Taken to the extreme case, this can lead to the activity occurring even without the stimulus, an occurrence called *vacuum activity*. LORENZ raised a starling in his house; he observed that "the bird looked concentratedly at the white ceiling of the room as if insects were flying there and then started flying, snapped at the air and returned to his place. It then made the movement of shaking its prey to death and swallowed. Then it was calm once again". The transfer of this perception to other behavioural patterns, especially to the *drive theory of aggression*, did not withstand scientific examination. The conditions under which animals are kept, for example, in zoos, are known to cause motion stereotypes. Under constant conditions, the motion is always carried out the same way over and over again so that a stiff and senseless motion pattern is apparent to the observer.

Displacement activity

In some conflict situations, behavioural patterns occur that seem unfitting and misplaced to the observer. Cocks of all chicken breeds stop fighting every now and again and start pecking, even though no food is available. When Eurasian oystercatchers see themselves in a mirror, they put their beak in their feathers and start to "sleep". Such seemingly senseless actions were interpreted as *displacement activities*, i.e. the mutual inhibition of stimuli of the same strength for two different behavioural patterns allows a third to occur. Newer models assume that the exhibited behaviour can indeed have a purpose if, for example, fight or reproduction success is increased by it (*fitness increase*).

Methods used in behavioural science

approaching

flapping of fins

shaking

cleaning

spawning

Behavioural patterns of a cichlid

Cichlids are popular fish for aquaria and are good models for observations of various behavioural patterns. The examination of single individuals and the objective analysis of the behavioural patterns that can be recognized and typed are the starting points of this approach. If all observed patterns of behaviour are noted in the order and frequency that they occur, then conclusions can be drawn about certain aspects of behaviour, e. g. courtship, brood care, feeding and territorial behaviour.

Such behavioural descriptions always have to be objective; they require the use of precise terminology and the separation of observation from interpretation. No anthropomorphism (humanization) or judgement is allowed. Thus, today, a fox would not be called a "sly fox", as is found in ALFRED BREHM's "Animal Lives" (1864).

Methods of molecular biology, such as genetic fingerprinting, help to discover relatedness in social organizations and thereby enable the interpretation of behavioural patterns. Technical equipment such as small transmitters or film and audiotape records make observation easier but also require a great deal of effort with respect to equipment and time. Directed experiments in the laboratory or under special conditions in a zoo can provide initial findings and can limit the research area. In contrast, some laboratory experiments only make sense when they are associated with observations in the wild. For example, when trying to breed bearded tits in captivity, an incomprehensible behavioural pattern occurred. After breeding successfully, the birds regularly pushed their offspring out of the nest. Since there was plenty of food, the offspring lay full and calm in the nest and did not show the normal gaping behaviour that indicates their need to be fed. In the wild, this apathy is a sign that the chick is sick or dead. However, when the parent birds received less food to feed to their chicks, they were then able to breed offspring successfully.

Experiments with <u>decoys</u> can reveal the single stimulus that is important to activate a behaviour. An artificial stimulus has mainly been used to analyse genetically determined behavioural patterns. However, learning processes cannot be excluded in these experiments and thus experiments carried out under the unnatural conditions might lead to false results. In the so-called Kaspar-Hauser experiments, animals were observed that could not learn a particular behaviour from fellow animals because they were kept in isolation. Under such conditions, behavioural disorders often appeared. Since not all animals always react in exactly the same way, statistical analyses are a normal part of such experiments nowadays, with statistical significance needing to be shown before a connection between a stimulus and reaction can be regarded as established.

Newer tests use often many different methods at the same time; these are focused more on physiological, ecological or evolutionary biological mechanisms depending on a causal or functional approach. A special problem is the study of human behaviour. Human ethologists who look for genetically determined behavioural patterns have to be able to exclude all cultural and social influences. Sociobiologists and behavioural ecologists need extremely large amounts of data for the examination of fitness consequences of human behaviour. This is also true for evolutionary psychology, which tries to explain behavioural patterns observed today based on adaptation processes in the past.

»info box«

Choice experiments

In this experimental set-up, we can test by which criteria cichlids choose their reproduction partner. The females spawned in 16 out of 20 experiments in nest B bottom left. The females watched the courtship display of both males and spawned.

Task

① Analyse the experimental set-up and explain the result by taking causal and functional aspects into consideration.

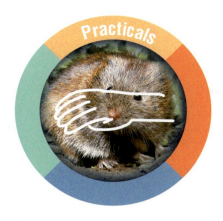

Wall-seeking behaviour in mice

Observations of mice in the wild suggest that they try to avoid crossing open spaces. This hypothesis was tested in experiments.

Experiment 1

Materials: plastic foil, a long ruler, overhead marker, cardboard, 1 stopwatch, one test mouse per group, writing pad with 8 · 8 squares

Preparation:
Paint a grid with 8 · 8 squares on the foil, each square should have an edge length of 10 cm. Place the foil on the table with the grid facing down, so that the animals are not distracted by the smell of the ink. Place strips of cardboard that are 20 cm high and covered in foil along the outer lines and connect them with adhesive tape.

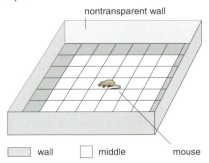

Experiment:
The mouse is placed inside the box and allowed to settle down (acclimatise) for 3 minutes. One group member should make a visible sign every 3 seconds. In these intervals the writer should note the position of the mouse on the writing pad for 5 minutes. The tip of the nose determines the position of the mouse. In order to remove odour marks, the foil should be wiped after each experiment.

Tasks

① Analyse the number of cases in which the mouse is in the squares that contact the wall (measuring value f_M).
② The expectancy value f_E (see statistical analysis) is made up by the number of grid squares for both areas (observe: 5 · 20 measured values).
③ Analyse the experiments by using the χ^2 test (see box).

Experiment 2

Possibly, a relatively frequent stay of the mice close to the wall area will be observed. This could have various reasons and should be examined by varying the experimental conditions.
— For this, the foil is placed on a board (80 · 80 cm) that is raised so that the margins are at least 70 cm above the ground. There is no outer boundary attached on two sides. A further change of the experimental set-up might be the introduction of a box that covers 4 squares in the middle. The analysis of these two experimental variations is done as in experiment 1.
— A further variation of the experimental set-up gives the following results:

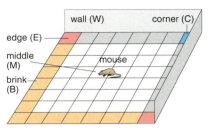

M	W	B	E	C
38	27	26	5	4

Tasks

④ Test with the χ^2 test whether the variations of the expected values are significant in this experiment.
⑤ What is the biological importance of the behaviour of the mice in both experimental set-ups? Describe the proximate and ultimate reasons, respectively.

Statistical analysis

The χ^2 test is a statistical test to determine whether a hypothesis applies with a certain probability or is false. In order to determine the probability (P), a table is used according to the example on the right. The χ^2 value is calculated from the expectancy value f_E together with the measured value f_M. The number of degrees of freedom (F) is the number of variables minus 1.

	f_E	f_M	$\frac{(f_M - f_E)^2}{f_E}$
wall			
middle			
χ^2 value			$\Sigma =$

Example: In experiment 1, the degree of freedom is F = 1, because there are two variables, the wall and the middle.

Degrees of freedom (F)	P							
	0.5	0.3	0.1	0.05	0.025	0.01	0.005	0.001
1	0.455	1.07	2.71	3.84	5.02	6.64	7.88	10.8
2	1.39	2.41	4.61	5.99	7.38	9.21	10.6	13.8
3	2.37	3.67	6.25	7.82	9.35	11.3	12.8	16.3
4	3.36	4.88	7.78	9.49	11.1	13.3	14.9	18.5
5	4.35	6.06	9.24	11.1	12.8	15.1	16.8	20.5

In the respective row of F, the value is sought that equals the calculated value or is just under it. In the upper column, the probability is found that the variation from the expected value is only coincidence. For a P ≤ 0.05, the variation is significant.

2 Behavioural patterns and their causes

Genetically determined elements of behaviour

We can assume that certain behavioural patterns are determined genetically if all individuals of one species or breed exhibit comparable behaviour. This is also the case if the behavioural elements occur without mistakes from birth and are stereotypic, which means that they always occur in the same way.

Mendelian features

The roundworm (*Rhabditis inermis*) lives in rotting material and faeces. It is not able to move to a new dunghill by itself but is distributed by a beetle. Its larvae attach under the wings of the beetle and are thus transported. Of this worm species, two subspecies exist that differ in the behaviour of the larvae. One type of larva waits motionlessly for random contact with a beetle. The other type sits on top of a dunghill while making swinging movements with the upper body and thus increases the probability of meeting an insect. If the "waving" roundworms are crossed with "non-waving" ones, the F_1 generation contains only waving offspring. If these offspring is crossed with each other, the "waving" and "non-waving" worms occur in a ratio of 3:1 in the F_2 generation

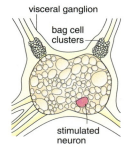

Genes for egg deposition

Sea hares (*Aplysia*) are sea slugs that are hermaphrodites and possess no shell. Here, more extensive evidence of a genetic base for behaviour has been found. Their reproductive behaviour contains fixed sequences of stereotypic behavioural patterns for which egg deposition has been especially well examined. After fertilization, Aplysia lays its eggs in long spawn strings that can have up to 1 million eggs in it. As soon as the muscles of the genital tract contract and the spawn string comes out of the genital opening, the animal stops creeping and eating. Heart beat and breathing frequency rise. The slug takes the beginning of the spawn string in its mouth and pulls it with characteristic swinging movements and thus helps the eggs to emerge. (The genital opening is located at the front of the body on the right.) Thereby, the spawn is wrapped into a ball and covered with a sticky secretion from a small slime gland in the mouth of the slug. Finally, the animal attaches the sticky spawn mass with a strong head movement to a stable base. The nervous system of Aplysia is quite primitive. During a search for the control mechanism of the described behaviour, two neuron groups were found, each arranged in a grape-like cluster (bag cell cluster) above the visceral ganglion. An extract from these cells injected into a slug induces the entire egg deposition behaviour even without previous mating. A peptide (egg-laying hormone, ELH) was isolated from this extract. It works via the blood like a hormone and activates breathing, heart beat and the muscles extruding the spawn string. As a neurotransmitter, it activates a specific neuron that is found in the visceral ganglion and that influences behaviour. The basis of the behaviour could thus be traced back to a DNA sequence. Gene-modifying experiments have identified two more *genes* for peptides that stimulate behavioural elements in Aplysia.

Genes and the environment

Genetically caused behaviour in animals can only be optimized over many generations and thus be adapted to the environment. However, only the reaction norm is determined in this way. In the spectrum of possibilities, each individual has to react

"Waving" roundworms

visceral ganglion
bag cell clusters
stimulated neuron

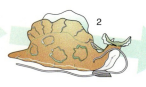

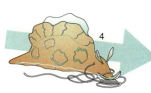

1 Behaviour during egg deposition of Aplysia

flexibly to current problems so that its reproductive success guarantees the passing on of its genes. That means that, in most cases, the genetically determined behaviour is modified by experience and is adapted to the environment.

Examples of this include several species of small African parrots that are known as lovebirds. One of them is the *rosy-faced lovebird*, which brings nesting material to its breeding cave by putting blades of grass and the cut parts of plants in its back feathers. The short lengths of plant material are kept in place by small hooks of the feathers. A closely related species (*Lilian's lovebird*) uses longer grass blades and carries them in its beak to the breeding cave. Both species can be crossed experimentally and their offspring show intermediate nesting behaviour. The females cut pieces of medium length and carry them in different ways: some put them in their feathers but do not let go of them, whereas others do not put them in correctly or just drop them. Eventually, the birds learn to carry the grass in their beaks but they always make a pretended attempt to put it between the feathers first.

Special environmental factors can change the behaviour of many generations. Mice use nesting materials in different amounts (fig. 2). In lab experiments, only animals that used a lot (line 1) or little (line 3) nesting material, respectively, were mated for numerous generations. As a control, random mating of the original populations was also examined (line 2). Even after only 15 generations, a measurable difference between the test groups and the control group could be seen. Whereas in these experiments, humans work as the selection pressure, ecological and physiological factors are crucial for the reproductive success of the examined animal species in the wild. Thus, they determine the transfer of behaviour to the next generation.

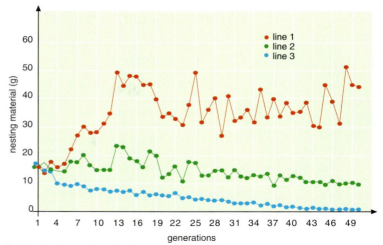

2 Nesting behaviour of mice

Tasks

① What conclusions can you draw from the crossing experiments with roundworms?
② In what way are the results of the crossing experiments different in the roundworms, mice and lovebirds?
③ Summarize the methods demonstrated here with which genetically determined behaviour can be shown.
④ Previously, genetically determined behavioural elements were called congenital. Compare the meaning of the two terms.

Rosy-faced lovebird with nesting material

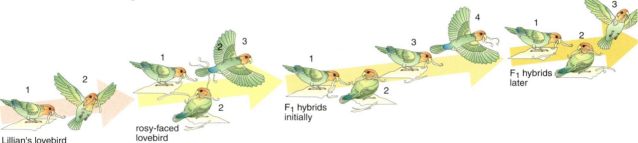

1 Behaviour of the rosy-faced lovebird

Behavioural Science

Internal and external motivators

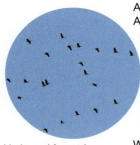

V-shaped formation of cranes during bird migration

Chronobiology
The field of science that examines the causes and functions of the periodic organisation of organisms.

A hungry animal goes on a quest for food. An animal ready for reproduction puts on a display or chooses a partner for mating. In nutritional, reproductive and brood-care behaviour, clear connections are known between internal conditions and external visible behaviour; these can be analysed, for example, via blood glucose levels or by determining hormone concentrations.

What trigger causes the blackbird to sing at 5 o'clock in the morning, the migration of cranes at a tightly restricted time of the year or the migrations of herds of hoofed animals? An examination of these questions within chronobiology also applies to humans. A cave researcher who isolated himself from the world for 3 months without a watch or any other time indicator was unable to state the correct date or time of day when he came back to the surface. If a person flies across several time zones, sleepiness during the daytime and sleepless nights, loss of appetite or digestion problems can occur (*jetlag*). In both cases, it takes several days until the internal timekeeper is readjusted to the conditions at the new location.

Like all higher organisms, humans have a biological clock that makes sure that their numerous physiological processes run with a rhythm of about 24 hours (= circadian rhythm).

Interactions between specific genes (e. g. *clock*, *period*, *frequency*, *timeless*), their mRNA and some proteins form the basis of such processes. So-called time-keeping proteins are produced in regulated amounts for specific metabolic processes by feedback loops. The activity rhythm of an organism is coupled to the natural day-night cycle by light exposure and other environmental contacts. This can be modulated by specific neurons of the serotonin system. The neuronal centres responsible for the control of the endogenous day rhythm are located in humans in a specific region between the hypothalamus and hypophysis (see page 41) and is connected on one side to the optic system and on the other side to the brainstem. The produced cyclic changes of body and brain function are able to control not only sleep and activity during the day, but also behaviour throughout the year in animals and in humans. The singing of the blackbird and the migrations of cranes or hoofed animals are thus triggered when certain internal conditions and the corresponding external time-keepers are in synchrony.

Time measurement by the circadian rhythm (*chronometry*) is the basis for the mechanisms of orientation that use the magnetic field, sun, moon or stars. For a specific behavioural pattern to occur, not only the internal conditions and the perception of the current position of the sun are necessary; animals must also take the continuously changing position of the sun into account by using an internal clock and additionally recognise its dependence on the latitude and the season. The yearly migration of the crane can be understood on this basis.

Tasks

① If people live for a few weeks in a room isolated from the outside, they develop the circadian rhythm shown in fig. 1. Explain the graph.
② After a transatlantic flight, many people complain about sleeping problems and they feel dull (jetlag). Explain this.
③ How can you distinguish between days of same lengths within the annual rhythm?

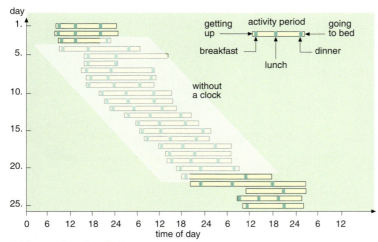

1 Human circadian rhythm

Behavioural Science

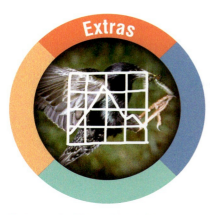

External stimuli driving motion

Biological long-term clocks control, for example, the periodically occurring migration of birds and the cycle of bodyweight including the accurately timed storage of fat. These clocks, like the basis of orientation behaviour, are determined by genetic factors. The internal conditions work together with external stimuli and make astonishing events possible, such as the migratory behaviour of the monarch butterfly or the long air routes of several 1,000 km undertaken by migrating birds.

Orientation movement

The non-directional movements of an animal are often changed by environmental factors such as light, heat or humidity (kinesis). In other cases (taxis), these factors can have a directing function. Both occur automatically. Migrating animals orient in the simplest case by known landmarks (e.g. coasts, mountains or the course of a river) in order to reach a destination far away (piloting).

Tasks

① Summarize the main differences between the three mentioned orientation movements.
② Relate the following behavioural patterns to the orientation movements and explain their biological importance:
— the activity of woodlice increases in dry environment and decreases in humid conditions.
— if the position of pot plants that are close to a beehive is experimentally changed it becomes more difficult for the worker bees to find their way home.
— fly larvae move away from the light after eating.

Orientation and navigation

If the animals are too far away from their destination to see it, they have to first determine their own location and then continuously control their motion in the direction of their destination (compass orientation and navigation). For this, the magnetic field can be used, for example, whose lines of magnetic flux have a defined direction and strength on every point of the earth. Additionally further reference systems that are either genetically determined or learnt (sun, moon, stars) can be employed. The compass can be used for various purposes:

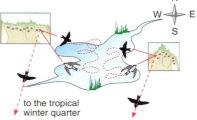

When Sand Martin birds of both colonies search for food, learnt navigation methods become important. When they migrate in autumn to their winter quarters they use a genetically determined direction formation that defines the destination course.

To resolve orientation mechanisms, 11,000 common starlings were examined that were on their way migrating from their breeding grounds to their winter quarters (see figure). The birds were transported from the Netherlands to Switzerland and released there.

Tasks

③ Interpret the two figures (see middle column).
④ Define the terms "compass orientation" and "navigation". Find out which animals navigate.

Compass in the eye

When changing location, the European robin orients by the magnetic field of the earth by using only the right eye. Its direction is influenced by photopigments that enter into an excited state on light absorption. From this state, they can reach an ever more excited state. This transition depends on the arrangement of the molecules relative to the magnetic field and can thus be used for orientation. To examine this, the following experimental set-up was used by Wolfgang Wiltschko. An aluminium funnel is covered with coated paper on which the bird leaves marks. The preferred activity direction can thus be identified. Such orientation cages can be lit with light of various spectral composition or, in an outside set-up, they can be placed into a coil system by which the magnetic field can be modified.

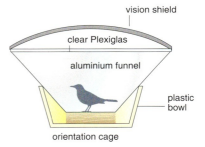

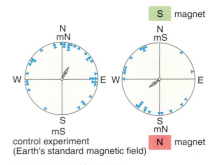

mN = magnetic North pole
mS = magnetic South pole

Task

⑤ Explain the experimental set-up and results.

Behavioural Science

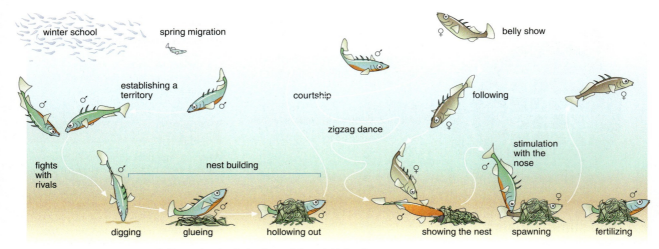

1 Reproduction behaviour of the stickleback

Behavioural sequences

The prey catching behaviour of the common toad (see page 68) is a classic example of genetically determined behaviour composed of appetency behaviour (instinctive desire), taxis (tactic movement) and the final action (consummatory act). Most behavioural patterns, however, consist of several elements that seem to be arranged like the links in a chain.

The water-boatman, an aquatic insect, localizes its prey by recognizing waves on the water surface. It is able to distinguish dead from alive objects by information it receives from its vibration-sensitive receptors. If an insect struggles on the water surface, the water-boatman swims towards it in a direction perpendicular to the the waves. Once it reaches its prey, it grabs it and holds it tightly. A metal fly, for example, would immediately be released. It stabs and sucks out the prey after checking it with its chemical sensory organs. The action sequence is thus determined by the features of the prey; taxis and the final action are mutually dependent.

The reproductive behaviour of sticklebacks (fig. 1) is a classic example of an action sequence in which two individuals of a species send signals to each other. This behaviour is triggered by external factors such as rising temperatures and an increase in the length of the day. Internal causations, such as rising testosterone levels in the blood, are also important in this case. The male stickleback establishes a territory by fights with rivals and then carries out nest building, courtship and brood care.

In contrast to earlier hypotheses, observations in the wild have revealed that the chain of actions does not break when one signal is absent. Wherever the action sequence is dictated by activating stimuli, links of the chains can be jumped over or, in certain cases, can even be repeated (fig. 2). Most reactions are triggered by several actions of the partner. The individual actions are always adapted to the reactions of the partner.

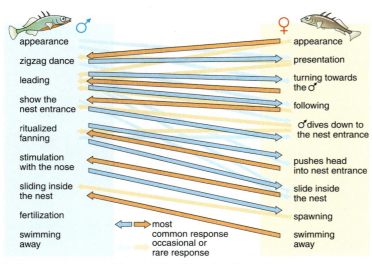

2 Action sequences of the stickleback's mating behaviour

Behavioural Science

The behaviour of the red-banded sand wasp

Brood care behaviour

Ammophila pubescens, which belongs to the family of thread-waisted wasps, overwinters as a permanent larva in a cocoon. The animals hatch in mid-May. After mating, the females show interesting brood care behaviour. Three phases can be clearly distinguished:

1. phase: The sand wasp digs a nest in the earth. After finishing the nest, the sand wasp closes it temporarily with pieces of sand and wood and then goes hunting for a caterpillar. The caterpillar is anaesthetized by a sting from the sand wasp and is transported into the nest. Here, the sand wasp lays one egg onto the caterpillar, leaves the nest and seals it. The paralyzed caterpillar serves as food for the hatched larva.

2. phase: Subsequently, the sand wasp visits the nest several times. This can occur without a caterpillar ("inspection visit") or the sand wasp delivers one to three more caterpillars to the hatched larva ("provision visit"). In the meantime, the nest is always re-sealed.

3. phase: In the last phase, the larva is supplied with up to 10 caterpillars before it becomes a pupa ("multiple-caterpillar day") and the nest is sealed for good. One sand wasp female can look after several nests during the same time period.

The behaviour of the thread-waisted wasps was mainly examined by GERARAD P. BAERENDS in 1940. The figure below shows a protocol of observations of the brood care of a sand wasp with 5 nests.

Tasks

① Describe the phase that can be seen for each nest.
② In what state were the nests on the evening of the 9th August?
③ List reasons for the assumption that the brood care behaviour is genetically determined.

The caterpillar as stimulus

"When a sand wasp hunts, a caterpillar is caught and stung. The caterpillar is pulled inside the nest if it lies close to the nest entrance just as the wasp has opened the nest. However, if the wasp was just going to close the nest, it sometimes uses the caterpillar as filling material. If one puts a caterpillar in the entrance during the nest building process, the wasp treats it as any other obstacle, e.g. like a piece of root, and tosses it aside." (quoting BAERENDS)

Task

④ Explain the differences in behaviour of the wasp concerning the caterpillar.

Putting the caterpillar inside the nest

The transport of the caterpillar occurs always in the same way. First, the prey is dropped immediately in front of the nest. The wasp then opens the nest entrance. When digging is finished, the wasp slides into the burrow and turns around without leaving the entrance completely (its abdomen remains in the nest entry). The wasp grabs the caterpillar and pulls it backwards inside the nest. While the wasp is "turning around", the caterpillar can be moved by the investigator so that the sand wasp has to leave the nest completely in order to grab the caterpillar. In this case, it once again drops the caterpillar close to the nest and pulls it inside backwards. BAERENDS was able to pull away the caterpillar about 20 times during the turning process before the sand wasp gave up and flew away.

Task

⑤ Compare the described behaviour of the sand wasp with the action sequence of the stickleback.

Interruption experiments

If, before an inspection visit, a young larva is replaced by an old larva that has just started to become a pupa, the sand wasp closes the nest for good. If the larva is replaced after the inspection, the pupa is treated according to the developmental stage of the removed larva. Thus, more caterpillars are delivered.

Task

⑥ Explain the importance of the inspection visit by using the interruption experiment.

Behavioural Science

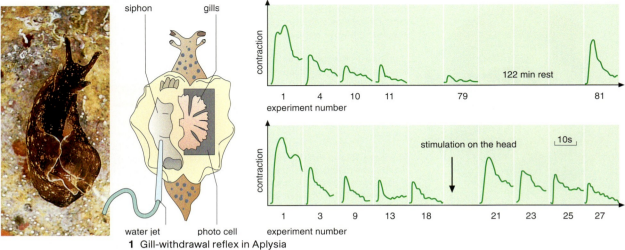

1 Gill-withdrawal reflex in Aplysia

Reflexes can be influenced

Habituation
Unconditional reflexes become weaker by stimulus-specific adaptation.

Dishabituation
Complete activation of a behavioural reaction, which had previously been reduced by habituation.

Reflexes represent a direct connection between stimulus and reaction. They are genetically determined and mainly occur in the same way. This is guaranteed by a reflex arc, which, in the simplest case, is made up of two nerve cells (see p. 24). They mediate between stimulus uptake and the reaction of a muscle or a gland.

The regulation of reflexes on a cellular basis can be readily examined by using the primitive nervous system of the sea slug *Aplysia*. If one strokes the back of the slug, it reacts by retracting its gills. This withdrawal reflex involves sensory cells in addition to sensory and motor neurons (fig. 2).

By using a light source and a photo cell that is illuminated more intensely when the gills are withdrawn, the strength of contraction can be measured (fig. 1). The results show that the strength of the reaction decreases continuously after repeatedly stimulating the back. This effect is called *habituation*. If the slug is then touched in another place, it reacts once again with the original strength (*dishabituation*). Thus, fatiguing of the organs responsible for the reaction can be excluded as a reason for habituation.

Extremely fine electrodes allow the excitation of the involved neurons to be recorded. This has shown that, whenever the strength of the stimulus remains the same, the motor neurons (adaptation phase) become less excitable after repeated stimulation, even though the stimulation forwarded by the sensory cells does not decrease (fig. 2). Neurophysiological investigations have provided evidence that the change of reaction strength of a reflex is made possible by additional nerve cells (*interneurons*). These are connected in parallel to the reflex arc and have an inhibiting or stimulating effect on the reflex centre. During habituation of the gill-withdrawal reflex, the impulses between the sensory neuron and motor neuron decrease, as do the impulses between the interneuron and motor neuron.

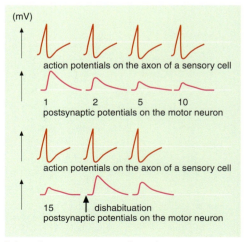

2 Impulses on sensory cells and motor neurons

Behavioural Science

Stick insect

More precise muscle coordination is needed for other genetically determined behavioural patterns that are more complex than reflexes and whose occurrence and intensity can additionally be influenced by the internal conditions of the animal. Stick insects react to weak blowing by slow swinging movements thereby resembling a branch in the wind. Stronger stimuli such as strong shaking lead to a form of stiffness (catalepsy). They can remain in this stiff position for up to 15 min. If a person carefully lifts one of its legs, the leg is moved extremely slowly to its original position as if in slow-motion.

Both behavioural patterns have the function of not disturbing the camouflage of the animal. Because of their branch-like body shape, posture and colouring, stick insects are extremely well adapted to their environment and any fast movements would reveal their presence to predators.

The animal reacts with the slow resetting reflex only if the external stimuli have signalled, to a higher nerve centre, that a possible danger exists. Through the connection via specific interneurons, the stimulating and inhibiting signals from the insect brain are transferred to the motor neurons of the leg muscles (fig. 1). The fast reaction of the flexor is inhibited. In any stage, blowing or careful touching can lower the readiness to fall into the stiff state. The animal immediately moves normally once again.

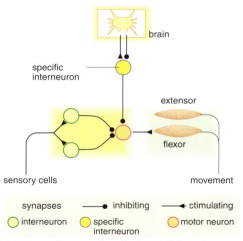

1 Neuronal connections in the stick insect

Tasks

1. Explain the measurement results of the repeated stimulation of Aplysia (fig. 78.1).
2. Compare figure 1 with the figure of the reflex arc (fig. 25.1).
3. If you move your head, the surroundings are still seen as fixed. If the eye ball is moved by pressing it lightly with a finger, the surroundings seem to move. Compare both actions and explain the differences.
4. If you walk around a market with a shopping basket, every weight change is automatically balanced. Explain.

»info box«

Walking with six legs

Stick insects only have about a few thousand nerve cells in total. Despite this, they are able to coordinate their 6 legs and 18 joints while walking and in any other body position when hanging on a branch. In the simplest case, they walk like all other insects: three legs are in the air and swing to the front, while the others are stable on the ground, support the body and push it frontwards. Each step cycle requires the control of several antagonistic muscles that are controlled by extensor and flexor neurons.

In the standing animal, the body posture remains the same because of the resistance reflex, even if a leg is lightly touched. Sensory cells located in the leg segments are stimulated by stretching. The motor neurons of the extensor fire and thus counteract a flexion.

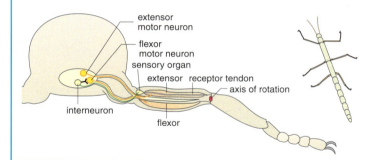

Behavioural Science

Conditioning — an animal as machine?

Conditioned reflex
This is caused by a stimulus that, by experience, has been connected to a reflex action.

Conditioned appetence
A conditioned stimulus is connected to an existing action.

Conditioned aversion
A neutral or previously positive stimulus becomes connected to defence or flight by bad experiences (punishment).

Conditioned action
Associations between a behavioural element and an active drive formed by learning.

Conditioned inhibition
Inhibition of a certain behaviour after receiving a negative experience while doing this behaviour.

Classical conditioning

In a dog, the secretion of saliva when seeing food (the *unconditioned response*) is a genetically fixed reaction. After a defined stimulus, it always occurs in the same way. If a previously neutral stimulus (sound of bell) occurs often enough together with the food, the sound becomes a conditioned stimulus that eventually causes salivation by itself. This *conditioned stimulus* can be erased once again if the sound is presented to the dog repeatedly without showing food. This is called *extinction* (deletion) in contrast to passively forgetting.

Based on these experiments of the Russian physiologist IWAN PETROWITSCH PAWLOW (1849 – 1936), the concept of *classical conditioning* was developed. It claims that all behavioural patterns are the result of conditioned reflexes.

In the experiments, the salivation of a tied-up dog was measured. A freely moving dog would walk to the bell, eat the presented food and hereby associate sound, reward and its action. In a similar fashion, other animal species learn flower features and scents when searching for food. Today, this learning process is thus called *conditioned appetence* because the change in behaviour is a result of positive experiences during an ongoing action. If a neutral situation is followed by a bad experience (e.g. punishment), this situation is avoided in the future.

Operant conditioning

In contrast to European behavioural scientists, American behaviourists considered an animal as black box. They claimed that it was only possible to measure stimuli and reactions, and that behaviour was composed almost entirely of learnt reactions. EDWARD L. THORNDIKE (1874 – 1949) and BURRHUS F. SKINNER (1904 – 1990) belonged to this group of behaviourists. Both examined learning behaviour with standardised experimental set-ups in which the animals were rewarded with food if they carried out "correct" actions: In THORNDIKE's problem box, a cat had to use a lever system in order to open the cage door and reach a reward located outside the cage. In the Skinner box, rats and doves learnt to push a lever or to peck against a panel in order to receive food.

Today, this behaviour is called *conditioned action*. The hungry animal shows exploratory curiosity, touches the lever by accident and receives a positive feedback (food). It does this action many times and constantly improves it, especially if it is hungry. If, in a similar manner, it learns not to do something because of punishment, this is referred to as *conditioned inhibition*.

Task

① Explain the differences between a conditioned reflex and operant conditioning.

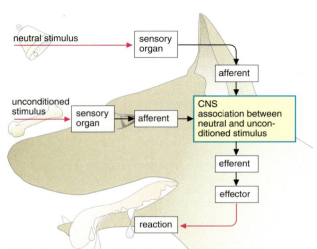

1 Functional diagram of classical conditioning

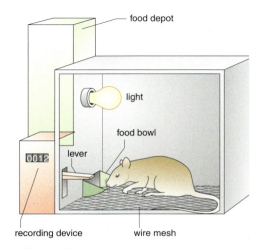

2 Rat in a Skinner box

Behavioural Science

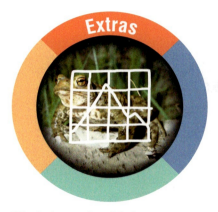

Models and criticisms thereof

Models are often simplified demonstrations that help the scientific understanding of structures, processes or mechanisms that thus become easier to comprehend and compare. Such understanding aids are mainly developed from analogies or similarities, i.e. conclusions are drawn from cases that resemble each other.

Functional diagrams

In the 1970s, BERNHARD HASSENSTEIN used the symbols of control technology for examples of behavioural science. For instinctive behaviour and the various ways of learning, he developed so-called functional diagrams that relate the units that are probably involved units and that represent the situation during learning.

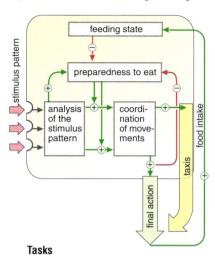

Tasks

① Describe the model for food intake and use it for the prey-catching behaviour of the common toad (see page 68).

② Explain why the model can describe prey-catching only incompletely.

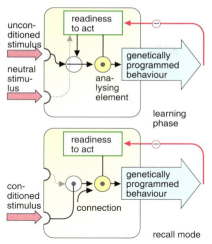

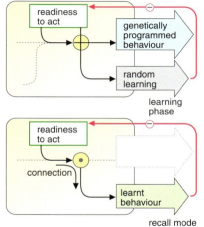

Tasks

③ Explain the functional diagrams of learning.
④ How do the two learning processes differ that are represented in the functional diagrams?
⑤ Find examples in which these easy learning processes can be found individually or in combination.
⑥ Draw conclusions about whether these models can be applied to observable behaviour.
⑦ Octopuses can be trained to swim out of their cave and to swim towards a white panel or to avoid it. The first behaviour occurs if the animals are fed during their learning phase and the latter if they are lightly punished. Analyse this by using the models.

Electronic reflex model

In an organism, reflexes are basic motor action units and are made up of a direct connection between a presynaptic sensory neuron and a postsynaptic motor neuron. Especially in vertebrates, this connection is embedded in complex connections of nerve cells whose cooperation during learning processes in the organism are difficult to analyse, even with electrophysiological methods. An early model for demonstrating a conditioned reflex in a simplified way can be built with an electronic kit:

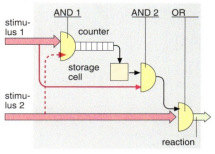

Explanation of the various building blocks:
— *Counter:* This is refilled by each incoming impulse and makes a contact after a certain number of impulses.
— *Storage cell:* If it is activated, it releases a continuous impulse.
— *AND-gate:* This connects further only if both inputs are occupied.
— *OR-gate:* This connects further if one or the other (or both) inputs are occupied.

Tasks

⑧ What happens in the model before learning, during learning and afterwards?
⑨ What can such a model be used for and which natural connections cannot be demonstrated with it?
⑩ Critics of classical conditioning assume that animals in such experimental situations do not react like automatons, but that they have learnt something about the relationship of events and react according to it.
Is this approach considered in the model? Explain your opinion and refer also to the Pawlow experiments.

Behavioural Science **81**

Learning and maturation

Many genetically determined behavioural elements are changed and complemented during the course of individual development. This cooperation between genetic and experience-dependent behaviour was called instinct-training interlacement by KONRAD LORENZ. Today, this term is rarely used because it is unclear.

Learning

In 1921, ornithologists (bird-watchers) in England observed blue tits that pecked open the metal-foil seals on the top of milk bottles and then sipped the cream layer. Obviously, one blue tit had opened a seal by coincidence pecking at it in the same way that blue tits usually peck at bark when searching for food. As the blue tit successfully also found food this way, it repeated pecking open shiny metal-foil seals. Learning enables the suitable adaptation of individual behaviour to special environmental conditions.

Learning processes take place in two phases: an organism takes in information during a stimulating situation and it saves this information in its memory (learning phase). In similar situations, the saved information is retrieved and causes — because of the experience — a changed behaviour (recall-mode). To drink the cream layer from a milk bottle is advantageous for the individual blue tit but it is not essential for survival. This is called f*acultative learning*.

Every squirrel has to learn an accurate opening technique for nuts (see box). Only then can they survive on their previously hidden food during the winter. In this case, learning is essential for survival and is thus called *obligatory learning*.

Not all animals of one species learn equally well. The worker bees of the Carniolan honey bee learn optical markings of their feeding places more easily than bees of the Italian subspecies. The reason for this are their different learning dispositions, i.e. their genetically determined learning abilities. Carniolan honey bees live in a region with unsettled weather and thus often need landmarks to aid orientation, in addition to orientation by using the sun compass. The bees of the Italian subspecies fly only in good weather so that orientation by the sun is sufficient. In contrast to individual adaptation of behaviour by learning, the various learning dispositions of species have developed during evolution during the course of many generations.

Blue tit

Learning
Learning is the storage of individually acquired information about the environment in a retrievable form within the memory.

»info box«

Modification of genetically determined reactions by learning

If a two-month-old and inexperienced squirrel *(Kaspar-Hauser animal)* is given a hazelnut for the first time, it starts gnawing immediately in all possible directions. Only later does it try to lever open the nut with its front teeth. Clay and wood beads that are the size of a nut are also gnawed on from all sides. The figure shows a gnawed on nut from its base (a) and tip (b).

A squirrel needs a lot of time to break open the first few nuts that it finds. The gnawing marks are, in the beginning, randomly distributed over the entire nut. Slowly after about the 12th nut, these marks become (c) aligned in parallel with the nut pattern. This causes a decrease in the time required before food intake occurs. Experienced squirrels immediately gnaw one or two longitudinal furrows or one hole in the tip or base before levering or breaking open the nut with their front teeth. Depending on the kind of used movements, three techniques can be distinguished in experienced squirrels: The bursting, the hole-gnawing and the hole-bursting technique (d, e, f).

Thus, every squirrel has the genetically determined ability to gnaw on nut-sized objects and additionally learns its own opening technique.

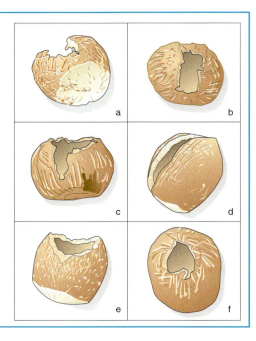

Behavioural Science

Scarecrow

Maturation
Development of a behavioural pattern based on a genetic background without the necessity of individual development (developmental process).

Conditioning
In contrast to p. 78, the simplest form of learning that not only involves unconditioned reflexes.

Sensitization
Simplest form of learning in which, after intense stimulation, the response increases to subsequent less intense stimuli.

Maturation

The perfection of behavioural patterns is not always based only on learning. An example is the flying ability of young doves. In an experiment, they were separated into two groups when freshly hatched. In one test group, the wings were fixed to the body and could not be moved. In the control group, the doves developed normally. When the control animals fledged, the flying ability of the test animals was also tested. They could also fly. Such development is called maturation. The full functionality of a behavioural pattern, in this case, is the result of developmental processes in the central nervous system and in the locomotor system.

Conditioning

In birds, we can see that a newly set up scarecrow will trigger flight behaviour. However, after a short while, the reaction to the constant stimulus decreases and finally stops. Another stimulus, e. g. a loud sound, can trigger flight behaviour at the original intensity. The observed decrease of reaction intensity is thus not explained by the fatigue of the involved muscles. We can also exclude a reduced readiness to flee. Since the scarecrow has neither positive nor negative consequences, the animal learns to adapt its behavioural response during the course of many associations with it.

Such a decrease of reaction intensity to a constantly re-occurring stimulus is called "conditioning". It is stimulus-specific. Experiments exclude adaptation of the sensory organs. Conditioning, which eventually disappears after the last stimulation, allows the avoidance of unnecessary reactions to constantly present stimuli.

Sensitization

Sensitization is a general rise in readiness to react. If, for example, sugar receptors on the legs of a fly are stimulated by a highly concentrated sugar solution as a chemical signal, the readiness to extend the proboscis increases for a few minutes. Diluted salt and sugar solutions are now sufficient as triggering signals. Extremely strong stimuli, such as mechanical blows, vibrations, light flashes or chemical signals, generally increase readiness to react to numerous other stimuli in a more intense and enduring manner.

Playing habits

Like behaviour in children, playing in animals is seen as behaviour that has no obvious reason but that contains many apparently directed movements. Young cats sneak up to their fellows and grab them with movements that are similar to the catching and killing of prey. However, they do not hurt their fellows, although the game still has risks and costs (increased energy expense). Two explanatory hypotheses can be proposed. According to the practice hypothesis, playing is a special kind of learning. The young animal can practice behavioural patterns "for later" and perfect them. The fitness hypothesis states that animals play when they are safe and well-supported in order to keep their muscles and cardiovascular system in shape.

Exploring

During exploring, diverse behavioural patterns are tried out on various objects and unknown habitats. Thereby, the animals come to know their environment without being rewarded or punished. Motivation and reward is part of performing the behaviour itself. This curious behaviour seems only to occur if social relationships are intact and elementary needs are met.

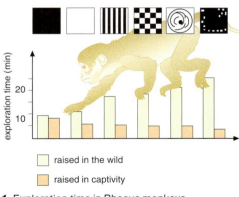

1 Exploration time in Rhesus monkeys

Behavioural Science

Imprinting

Goslings that have been raised in isolation and only have contact with humans behave really striking when they come into contact with other geese. Even when their natural parent animals approach them, they run away and, by their peeping sounds, they show that they feel lonely. However, the goslings follow every human who is close by. Chickens, ducks and some other birds and a few fish and mammal species act similarly.

This following reaction was termed by Douglas Spalding as early as 1875 as *imprinting* and was later examined in more detail by Oskar Heinroth, Eckhard H. Hess and, especially, Konrad Lorenz. Animals raised in an incubator were placed into a carousel containing a decoy that was equipped with a loudspeaker and that could be moved in a circle. The decoy could resemble a goose or duck but might also be a soccer ball. As early as the second day, the chicks followed the only decoy that they had seen on the first day. The chicks learnt the individual pattern only at a specific time after birth and preferred their imprinted object in choice experiments.

The characteristics of following imprinting are thus: the learning process takes place quickly and, during a *sensitive phase*, is mainly *irreversible* and highly effective. The behavioural pattern is genetically programmed and belongs to *obligatory learning*. In nature, accurate recognition of the mother is a prerequisite for following her onto the pond or river in the company of many fellow animals. This is vital for the chicks, as they are led by their mother to food and she defends them against enemies. In evolution biology, imprinting is known as device for identifying related animals of the same species. Every individual can thus adapt its behaviour towards the others depending on the degree of relationship.

Examination of the involved brain structures has provided evidence that imprinting mostly has a neuronal basis. In chicks, this occurs accompanied by microscopically and metabolic physiologically detectable changes of the synapses in a specific region.

In the described imprinting, a triggering stimulus pattern is learnt but this is not the case during the imprinting of birdsongs. Males of the chaffinch were isolated and raised acoustically shielded. If the songs of different species are played to them before they have developed their own song, their future song is determined by this experience.

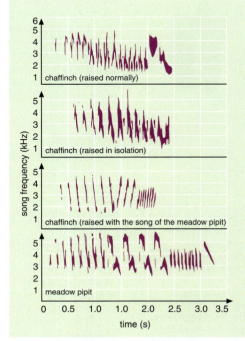

1 Song imprinting

If Zebra finches (nestlings) are raised by seagulls, then the adult finch male displays courtship only to the female of the foster parents. Sexual imprinting occurs during a time when the finches are not yet fertile. This demonstrates that the learning process and the activity related to the imprinting can occur a long time from each other.

Similar events are true for other kinds of imprinting such as *location imprinting* (e.g. in salmon), *habitat imprinting* of a few singing birds, *food imprinting* or *imprinting of breeding parasites* (e.g. cuckoo).

Task

① Analyse the experiments on bird song imprinting and compare it to other kinds of imprinting.

opening of loudspeaker
Imprinting carousel

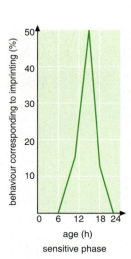

Sensitive phase
The time in which an animal is especially sensitive to certain experiences or at which it has to undertake certain experiences.

Shrew "caravan" formation

HANNA MARIA ZIPPELIUS has described that young house mice and shrews are encouraged by their mother to form "caravans" if they move out of the nest or have to change shelter. The mother demands "caravan" formation by their whining young in a graded intensity depending on the situation. The mother bites the young in its neck fur and briefly lifts it up (see figure).

The young bite either into the fur of the mother close to its tail root or into the fur of another sibling, the last one in the "caravan". The mother starts walking with the "caravan" of young following it (see figure to the right).

Decoy and exchange experiments

The formation of such a "caravan" was assumed to require communication between the involved animals. However, researchers wanted to know whether the required behaviour for "caravan" formation was genetically determined in the young or learnt by experience. The following observations give more information:

— If a young shrew is raised in isolation and by a foster mother of a different species that does not initiate "caravan" formation, it will bite into the fur of the foster mother or a nest sibling from the 6th day after birth.

— If an 8- to 14-day-old greater white-toothed shrew, which does not react to non-species decoys even after the strongest stimulation, is allowed to smell a living greater white-toothed shrew, it starts its familiar whining and will bite into all offered decoys.

— If at this time a greater white-toothed shrew is presented whose smell is altered by essential oils, it is not recognized and does not function as an object to stimulate biting.

— Between the 8th and 14th day after birth, the greater white-toothed shrew turns spontaneously away from all non-species decoys. It accepts it, however, if the species-specific smell was presented previously.

— From the 15th day after birth, a shrew bites only into the mother or a sibling; foreign decoys are threatened.

— If an animal of 4—5 days of age is given to a foster mother, e. g. a house mouse, the young shrew shows the biting behaviour from the 6th day after stimulation even when presented with different decoys. After the 10th day, this reaction will occur towards a dead house mouse but not towards the shrew mother.

— From the 15th day, the young shrew bites only into the foster mother and its stepsiblings but not into other house mice or its shrew mother and siblings.

— If a young shrew that was raised by a house mouse foster mother is given back to its real mother, it suckles again. The young does not react to the mother's stimulation any longer but immediately bites into the fur of the foster mother when it is returned to its nest. From the 21st day, when biting and walking together normally disappears, nothing further can change this behaviour.

Tasks

① Describe how the reaction to the animation to form "caravans" develops in growing up shrews.
② Conclude from the experiments to what extend the "caravan" formation of shrews can be described as learning by imprinting.
③ Explain the biological importance of the described behaviour.

1 Orang-utan in a box experiment

2 Japanese Macaque washing their food

Complex learning

In the past, it was assumed that the ability for complex learning was restricted to humans. Meanwhile, a range of examples of such abilities is also known from animals.

Tool usage

Chimpanzees were observed in the wild when they were scratching open a termite hill with their fingers. They looked around for a stiff stem or branch and shaped it with their teeth. Then, they held the stem into the hole until the termites had bitten into it. They removed the tool and ate the termites. Others dip chewed leaves into water-filled knotholes and then drink from it by sucking out the liquid. It is striking that the animals only show these techniques if they face a suitable problem. This is in contrast to cases of tool usage with genetically determined dispositions. The woodpecker finch uses stems to reach insect larvae underneath the bark. Young animals also use stems, even if the larvae are openly approachable.

Chimpanzee poking for termites

Learning by insight

WOLFGANG KÖHLER described, in 1917, an experiment with chimpanzees that were in a room in which a banana was hanging from the ceiling placed so high up that they could not reach it. A wooden box that had one open side was also in the room. The animals first tried to jump up to reach the target. One chimpanzee gave up rather quickly, walked around restlessly and suddenly stopped in front of the box. It took the box, pushed it in a straight line towards the target and hopped on top of it when it was about 0.5 m far away. While jumping, it grabbed the banana. The whole process took only a few seconds. The animal had solved the problem. It seemed to have tried out the solution to the problem in its thoughts before it carried it out. This is called learning by insight or newly combined behaviour (fig. 1). An essential criterion hereby is not the complexity of the problem but that the solution is found without previous experience and not by trial and error. As external features, the planning and action phases are clearly distinguishable from each other and are often separated by another phase. The action occurs quickly and is directed.

Imitation and tradition

Japanese Macaques have been observed for a long period of time on one of the Japanese islands. In 1953, a young animal was first seen to wash batatas (sweet potato) in a river in order to remove sand from it. Initially, only animals of the same age group imitated this behaviour but later almost all animals of the group washed their batatas before eating them. Animals of the following generations also expressed this behaviour (fig. 2). Learning by *imitation* leads to the creation of a *tradition*.

Other learning processes

Planned actions

In order to check whether apes can handle long action sequences including tool usage, a sequence of small boxes with various locking mechanisms were constructed. Each of them contained the opener for the next box. Which opener was in the box and which box contained the reward was visible to Julia, an orang-utan female, because the boxes had glass lids. She was trained in the handling of the individual openers. In a few cases, she worked this out herself by trial and error. In the experiments, the position of the boxes, the openers and the rewards were changed. At first, Julia examined the content of the boxes for a while, before she opened the boxes one after another and reached the reward without previous trial and error. She had understood the causative connection between cause and effect. This is an example of planned actions.

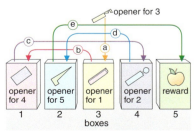

Abstraction and generalization

In order to test the ability of rats to distinguish between objects, they were offered three flaps with different stripe patterns. Two patterns were the same, whereas the third one was different. If the rat pushed the flap with the different pattern, it was rewarded. If the experiment was repeated with other patterns but the same situation of choices, the rat mainly chose the different pattern because it had been re-enforced.

The behaviour of the rat was interpreted such that the rat was not conditioned to a certain pattern but to the feature "different from the multiple occurring patterns" (*abstraction*). It also used this principle of choice behaviour in further experiments (*generalization*). The common feature "different pattern" had become a concept in the form of a picture or scheme (non-verbal concept formation).

Non-verbal and verbal concepts

OTTO KOEHLER offered several food bowls, each of which was closed with a lid, to an *alpine chough*. Each lid had a different number of dots on it. The bird learnt to lift the lid of the bowl whose lid had the same number of dots as on a sign shown to it. Only this bowl contained seeds as the food reward. The birds could recognize up to 8 dots as being the quantity shown in the picture (non-verbal numeric concept). Verbal concept formation is reserved to humans. They can express concepts as a thought process and in words (e.g. numbers).

Abstract value concepts

Chimpanzees and Rhesus monkeys received differently coloured game badges when they fulfilled a task. They could exchange these later: for a blue badge, they received raisins from a food automat; with another badge type, they could open the door of an enclosure; for a third badge type, they could play with the care-giver. The animals fulfilled tasks and collected badges. A female chimpanzee used a badge to open the door in order to flee from a cameraman. This example demonstrates that primates are able to generate abstract value concepts and use them correctly depending on the situation.

Self-recognition

If chimpanzees have the possibility of seeing themselves in a mirror, they start manipulating and removing colour spots that were put on them when they were asleep. The chimpanzees thus recognize the mirror image as an image of themselves. This behaviour shows that apes have a non-verbal self-concept. The question now arises as to whether they also have consciousness in the human sense.

3 Ecology and Behaviour

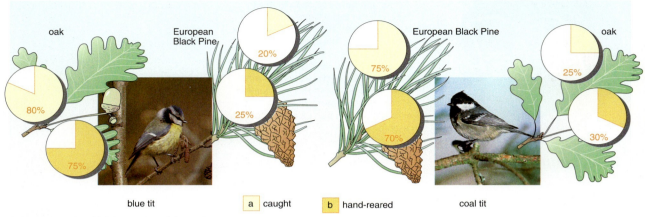

| blue tit | a caught | b hand-reared | coal tit |

1 Experiments with blue and coal tits

Habitat choice and territory

In field guides to birds, we not only find pictures of the individual species but also further details about them. These include their geographic range and typical _habitat_. We are thus given the information that _blue tits_ are common in mixed woodlands, whereas the _coal tit_ is mainly present in coniferous woodland.

Some scientists are interested in the reasons that animals chose particular habitats in which to live (_habitat choice_). Possible hypotheses are:
— animals migrate through various habitats and stay in those in which they find food most effectively.
— they also later prefer the habitat in which they grew up.
— they recognize the correct habitat on the basis of specific features and prefer this habitat as a result of genetic determination.

In order to clarify these connections, a researcher offered blue tits and coal tits an equal mixture of branches from conifers and from trees with leaves (choice experiment) within aviaries. She then checked with a stopwatch the length of time that the animals remained on the various branches. She performed this experiment with birds caught in the wild and with hand-reared birds. In both test groups, the blue tits preferred the branches with leaves; the coal tits, however, preferred the coniferous

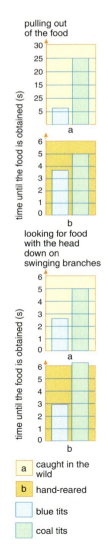

pulling out of the food

looking for food with the head down on swinging branches

a caught in the wild
b hand-reared
blue tits
coal tits

branches. Experiments with other species showed that these preferences can be enhanced by early experience with the preferred material.

Blue tits are often observed to hang from leaves when searching for food. They also pull on bunches of leaves and peck through leaves. Such techniques are seldom seen in coal tits, which look for food on the needle-like leaves and the bark of conifers. In experiments, scientists offered food hidden between leaves to blue tits and coal tits that were caught in the wild or hand-reared. The time taken for the birds to find their food was measured. In all cases, the coal tits needed more time to find the prey between the leaves and to catch it (see figure in the middle column). The above two experiments show that blue tits prefer the habitat to which they are better adapted than the coal tits. The blue tits are superior to coal tits under these conditions.

Whether the animals can actually settle in their preferred habitat depends on various factors, e.g. the presence of predators and of birds of the same species.

In early spring, we can often see blackbirds that hop side by side with spread feathers and a lowered tail until, all of a sudden, they fly up in front of each other while flapping their wings and then they attack each other. This all happens as if directed by unseen forces. If the locations of such fights are noted on a map, we find that each bird defends a well-defined area. Ornithologists made similar observations many years ago and called the defended area the bird's *territory*. Since territorial behaviour has been observed in many bird species, it is thought of as a common biological phenomenon. Some mammals are also known to be territorial.

At certain periods of time, some animal species are not inclined to defend an area. They use large areas peacefully together with other members of the same species and drive others away only from small core areas, which, however, do not contain all the vital resources that they need. Thus, in addition to territories, animals of the same species frequently co-exist peacefully in overlapping action areas. The defending of geographic areas often only occurs at specific periods of time. This has lead to the conclusion that the reason for the formation of territories is not generally to secure resources.

Today, we know that the basis of female territorial behaviour is mainly the competition for food. Females need optimal food resources for raising their young, i.e. to secure their own reproductive fitness. Males, in contrast, can improve their fitness

territories of pairs

harem territories

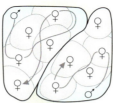
subterritories

territorial patches of males

by having access to several females. If the competition for females is higher than for food, the males show especially pronounced territorial behaviour. Depending on ecological conditions, females have either larger or smaller territories than males or are not territorial at all. Males usually defend larger areas depending on the distribution of females. Because of overlaps in space and time, these territories can have complicated patterns (see column in the middle).

Whether, during a certain period of time, an individual acts territorially depends on several factors. The defence of a territory follows a cost-benefit calculation. Figure 2 shows the calculation of the optimal territory size.

In a few animal species, so-called *floaters* exist. They stay in the territory of species members without attracting attention and avoid conflicts with the owner of the territory. These are usually males that take over the territory of their predecessors if the latter disappear for some reason. In great tits, males and even female floaters have been found. They can form couples and secretly breed in the margin of a foreign territory but with low reproductive success.

Animals without territories are usually pushed into areas with poor food resources and higher predator pressure and thus more often fall victim to a predator.

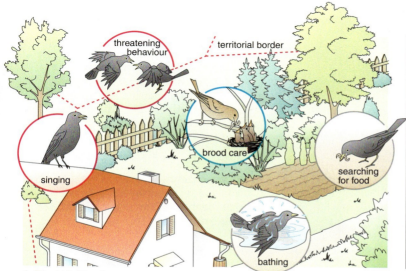
1 Blackbird territory

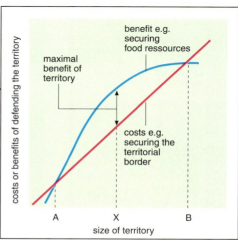
2 Conditions deciding size of territory

Behavioural Science

Optimized nutritional strategies

Animals constantly use energy in order to maintain their body functions. This energy has to be restored by food intake, which comprises searching for food, breaking it down, swallowing and digesting it. All this requires energy and time. According to the theory of optimal nutritional strategies, animals have evolved in such a way that they unconsciously carry out cost-benefit analyses in order to maximize food intake.

The first question that arises is what kind of food the animal should choose. The biologists PETER JARMAN and DOROTHY BELL examined the food spectra of even-toed ungulates in the Serengeti and found a connection between body size and food quality (fig. 1).

Compared with their body volume, small homoiothermic (endothermic) animals have a relatively large surface area from which they release energy in the form of heat. Because of their relatively small surface area, larger specimen lose less energy per gram body weight and time. Smaller individuals thus release more energy per unit time than large animals. Released energy has to be replaced by food intake. This means that small animals with a small gastrointestinal tract have to convert relatively more energy than large animals. Thus, the food has to pass through the intestine faster and also has to be digested faster. The connection is summarized in the *Jarman-Bell principle*: small animals need food that is rich in energy and easy to digest.

The break-down of cellulose, the main component of plant cell walls, is a major problem for birds and mammals, since they do not posses the relevant enzymes to degrade it. Plant cell walls deliver roughage (dietary fibre) that are only slowly degradable by symbiotic unicellular organisms. Cells in buds or young leaves have thin cell walls and are thus easier to digest since they contain relatively little roughage. The same is true for flowers, fruits and, in particular, animal food in which cell walls are missing. The usability and quality of food increase from old cells via young cells, leaf buds, flowers and fruits, up to animal-based food.

Larger animals can thus commonly eat plant parts that are difficult to digest, whereas smaller species select and choose those plant parts of high quality. This has been also established for various primate species and for ungulates.

Task

① Summarize the facts shown in the figure and suggest how the nutrition of the European Roe Deer and the Red Deer might differ.

ⓐ flowers, tips of branches, fruits, seeds, sometimes meat

ⓑ some leaves and grass species that are eaten selectively

ⓒ several shrub and grass species, less selective

ⓓ many leaf and grass species in particularly early stages of development

ⓔ many leaf and grass species, also those of lower nutritional value

1 Nutrition of African ungulates (hoofed animal) depending on body weight

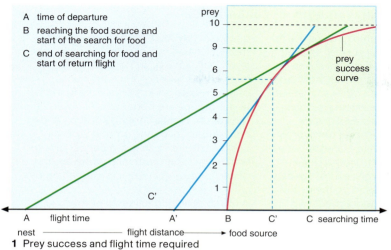

A time of departure
B reaching the food source and start of the search for food
C end of searching for food and start of return flight

prey success curve

1 Prey success and flight time required

2 Atlantic Puffin with prey

Searching for food

Seabirds such as the Atlantic Puffin catch their food (small sand eels) in waters close to the coast. During one dive, they can catch several fish, which they hold in their beak with their tongue. The more fish they have caught, the more difficult it becomes to catch more fish. This is demonstrated in the prey yield graph that starts at B, rises steeply and then levels out. The upper limit is 10 animals. During the time that an Atlantic Puffin is feeding its young, it has to fly from its young to the food sources and back again. Nestlings need approximately as much food as their own body weight; this means that the burden for the parents becomes greater and greater as the chicks grow. A single roundtrip consists of the time to fly there, the time to search and the time to fly back. In figure 1, success at catching prey is shown in relationship to the searching time and the time needed to fly to the food. At A, flight begins, whereas at B, the animal reaches the food source and starts to search for food. A bird achieves the maximum food quantity per unit time if the Y value (prey success) is as high as possible and, at the same time, the X value (time required) is as small as possible. In this case, the straight line between the time point when the bird flies off and the optimal food quantity is especially steep.

For a given starting point, this optimal "prey success-time relationship point" can be found by aligning a tangent with the prey success curve. This shows: the Atlantic Puffin must bring back more prey animals

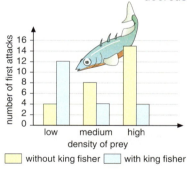

from food sources further away. Observations confirm this prognosis.

Searching for food and predator pressure

Grazing animals, e. g. the European roe deer, frequently stop eating and observe their surroundings; they safeguard themselves. The frequency of safeguarding decreases if the animal is in a group of the same species. MILINSKI and HELLER examined sticklebacks to which they offered swarms of water fleas of various density: hungry sticklebacks clearly preferred the denser swarms because here they could eat many prey animals within a short time. If the animals were partially satiated, they chose the less dense swarms. In a further experiment, the researchers placed an imitation kingfisher above the aquarium. The sticklebacks changed their behaviour and preferred regions of low prey concentration (see figure, middle column). MILINSKI and HELLER thus showed that catching water fleas in dense swarms required so much concentration that the fish could less well observe the predator at the same time. The behaviour shown is therefore a compromise.

Task

① Use the text to explain why it is favourable for the Atlantic Puffin to nest as close as possible to the coast.

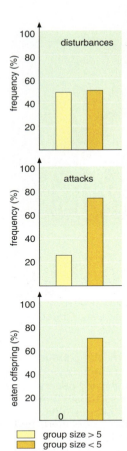

Advantages and disadvantages of cohabitation

Whereas individuals of some species live as solitary animals, others form small or large groups, at least temporarily. What advantages and disadvantages does an individual gain by living together with other animals of its species? Researchers who have observed the *dwarf mongoose* in East Africa provide an answer. These small carnivores of the family Viverridae have many predators that attack them not only from the sky, but also from the ground. Larger animal species have fewer predators. Based on the rule of "the smaller you are, the more predators you have to fear", the dwarf mongoose has to catch many animals every day but without becoming prey itself. Most attacks are carried out by birds of prey from which the mongoose can flee into its burrow.

The mongoose searches for its prey in high grass so that it is difficult for it to ensure its safety. Watching out to avoid being caught takes important time in which the mongoose cannot search for food. Thus, in most groups, a guard watches out while the others are eating. After a certain time, the guard is changed. Groups of fewer than 5 or more than 5 animals are disturbed to the same extent by birds of prey. Small groups, however, are actually attacked almost three times more often and lose thereby about $2/3$ of all offspring within the first 4 months after birth (see margin). All small groups (five and less) were wiped out within 2 years because they were not able to protect themselves effectively. Usually males of lower rank act as guards. If these are missing in smaller groups, the success rate of the predators increases. Furthermore, larger groups are better able to defend their territory from neighbouring groups and to fight predators on the ground such as snakes. However, in these denser groups with many social contacts, parasites and disease-causing germs can spread easily.

A group of fleeing animals initially runs in confusion but the animals quickly move closer together. This provides an advantage for the individual. The attacker cannot focus on one single animal (*confusion effect*) and the probability that an animal becomes prey decreases.

The protection from predators is the most important advantage of cohabitation, whereas the largest disadvantage is probably competition for resources. The intensity of competition pressure depends on a complicated interaction between body size, food quality and food supply. Hereby, the spatial and temporal distribution of the food is also of importance.

Dwarf mongoose at its burrow

1 Zebra and gnu herds together at a water hole

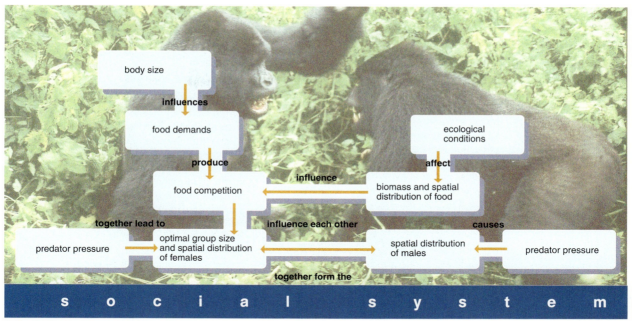

1 Ecological factors determine social systems

Social systems

Some animals live together in groups, whereas others meet up without setting out to look for each other. Elephant faeces lying in the African savannah contain many different beetles that fight for the faeces as food. Water holes are meeting points for zebras, gnus, rhinos and many other animals. Behavioural scientists call such meetings *aggregations*. They are characterized by the fact that the individual animals are looking for a location and not for fellow animals. Fish swarms, in contrast, form because a fish of one species looks for more fish of the same species nearby. Since an individual fish can join and leave the swarm whenever it wants to and since the fish do not know each other, this is called an *open anonymous society*. All members of a bee colony recognize each other by their hive smell but they do not recognize each other individually. Intruders are warded off. This social system is called a *closed anonymous society*. In *closed non-anonymous societies*, which often occur in birds and mammals, the group members know the other individuals of the group.

The population of a species is often composed of various groups including rogues males, bachelor groups, harem groups, and family groups. The combination in which males and females live together is called the social system. Four combinations are possible: *monogamy*, *polygyny*, *polyandry* and *polygynandry* (see middle column). Hereby, the social system is not determined within one species. Wood Pigeons, for example, can live monogamously during the reproductive period but, in winter, they migrate in polygynandrous groups. For some monkey species different social systems can be advantageous in different ecosystems. For a long time, the formation of such systems was not understood. Obviously, the social system is based on the different distribution patterns of females and males. New findings have revealed the following causes: competition for food and predator pressure are essential with respect to the optimal group size for females, whereas for the males, the competition for females is more important than competition for food.

Food requirements of females vary depending on body size (*Jarman-Bell principle*). For females, competition for food results from its supply and distribution. This, together with predator pressure, influences group size. The spatial distribution of females and males affects the social system, which changes with alterations in the basic ecological conditions.

Social systems

Monogamy
Social pairing of one male and one female for one reproductive season or until one partner dies.

Polygamy
One individual of one gender lives together with more than one partner of the other gender. Three forms are distinguished:

Polygyny
One male lives together with several females.

Polyandry
One female lives together with several males.

Polygynandry
Several males and females form a group together.

Behavioural Science

4 Evolution and Behaviour

Successful reproduction

If the main reason for reproduction is to maintain the species (as has been thought for a long time), then it would suffice if all members of a species reproduced at the same rate. Following the discovery of evolution, we now know that only those that reproduce more successfully than others win the "game of evolution". A comparison between individuals is possible if, after their death, the total number of their surviving offspring, and thus their lifetime reproductive success (their so-called *fitness*), is known. The connection between social behaviour and fitness has been examined in great detail in some species.

Case study: elephant seals

Elephant seals are superbly adapted to life in water. Only a few predator-free coastal areas are available for the females to use as breeding beaches; these occur mainly on islands. The seals give birth within the first 6 days after their arrival at the breeding beach. Subsequently, they suckle their offspring for 28 days with milk that is highly rich in fat. During their last 4 days on the beach, the females become ready to mate once again.

Reproductive behaviour

Elephant seal bulls, which are about three-times as heavy as the females, arrive at the breeding sites before the females and fight each other in order to establish a *hierarchy*. Success in these fights requires an enormous body size and substantial fighting force, both of which are reached by the animals only late in life. At the age of 8 years, males are ready for mating but they only really become successful when they reach about 10 to 11 years of age. By 13 years of age, they hardly win any fights. The maximum life expectancy is 14 years.

Younger sexually mature males avoid dangerous fights. The successful tactic for the young males is thus to wait. In contrast, a male of 10 to 11 years possibly has his last chance to reproduce. Older males pin everything on one chance. Such opponents rear up in front of each other, hit the rival seal with the chest and head, and bite each other with their canine teeth causing deep bloody wounds (*injurious fight*).

The females reach a maximum of 14 years. They can give birth to their first young at the age of 3 years but usually do not do so until the age of 4 to 5. Subsequently, they have one young per year and, thus, if they survive long enough, have a maximum of ten young overall. All females that reach sexual

Sociobiology
Sociobiology is the science of the biological adaptation of social behaviour. It measures the degree of adaptation of certain behavioural patterns on the basis of the lifetime reproductive success of the respective individual.

1 Elephant seals

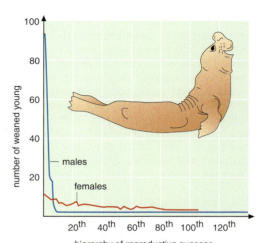

2 Reproductive success of elephant seals

1 Blue tit

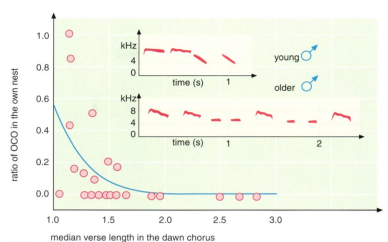
2 Out-of-Couple-Offspring (OCO) and song length

maturity have offspring, whereas many males cannot reproduce because the males higher in hierarchy stop them.

Choice of partner and gender conflicts

Decades prior to the investigations of elephant seals, the English researcher ANGUS JOHN BATEMAN discovered similar relationships amongst other species, although the importance of his results was realized by many researchers only decades later. As early as 1948, BATEMAN showed, in experiments with *Drosophila*, that an asymmetry exists between the sexes. Males can increase their fitness directly by copulating with many females; however, females cannot (see margin). BATEMAN concluded that, from an evolutionary biological viewpoint, males seek access to as many females as possible. Since the fitness of the female mainly depends on her reproductive success and the survival of her offspring, BATEMAN suggested that the females mainly set value on the "quality" of their mating partners, i. e. on "good genes". Signs of the good survival quality of the males include longevity and good health.

In many animal species, such as monogamous songbirds, conflicts occur often between partners. A conflict develops if the behaviour of one partner increases its own fitness but decreases the fitness of the other. This, for example, is the case in blue tits that breed as a monogamous couple in which both partners feed the young. With the help of the male, the female raises about 7.5 young, whereas if the male does not help, only 5.4 are raised. Females thus depend on male support. However, if during the partner choice in spring, the females end up with a less attractive male, then the other males in the neighbourhood possess "better" genes. Females can, in this case, raise the survival success of the own offspring by obtaining genes from an attractive neighbour, i. e. they cheat on their actual partner. This would increase the own fitness but lower the fitness of their partner. Blue tit males, especially the less attractive ones, watch their partner intensively during the fertile phase just before egg laying.

DNA tests (genetic fingerprinting) have established that the offspring of 10 — 15 % of blue tit females are not from their own partner, but from males from neighbouring territories. Hereby, any neighbouring females that are also unfaithful almost always prefer the same male, which is obviously particularly attractive. His own partner is not normally unfaithful.

The length of the song of blue tit males increases with age and experience (fig. 2). The "worse" the song of her own male, the more probable is it that she cheats on him with the neighbouring males that indicate their superior ability to survive by their longer songs.

Task

① Find reasons for the decisions made by the blue tit female from the viewpoint of a sociobiologist.

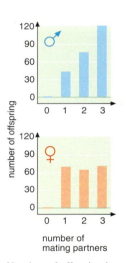
Number of offspring in *Drosophila*

Bateman principle
With each additional mating partner, the reproductive success of the males, but not of the females, increases.

Behavioural Science **95**

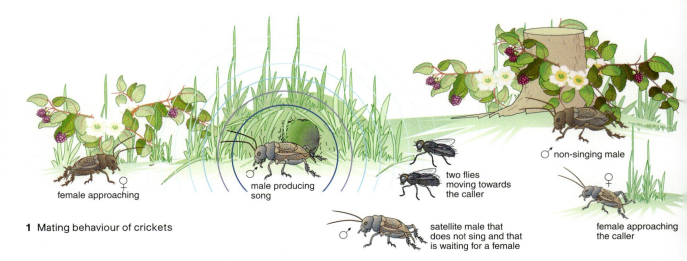

1 Mating behaviour of crickets

Strategies of sexual behaviour

Satellite male
A male that waits close to a courtship-displaying male and that tries to catch females that are ready to mate and that were attracted by the courtship-displaying male

Strategy
Genetically determined behaviour of an individual

Tactic
Behavioural alternative within a strategy

Behavioural polymorphism
Is present if different reactions with the same function can occur in response to a particular stimulus

Examination of courtship behaviour in crickets has shown that the song of a male cricket played by a loudspeaker attracts, with increasing volume, not only more females, but also non-singing males, so-called *satellite males*, and parasitic flies. The satellite males wait close to the singing male and try to catch females that are ready for mating. The flies usually lay their larvae on singing cricket males that are then slowly eaten from the inside by the invasive larvae. Satellite males are less often affected.

Breeding experiments have shown that the number of hours that a male calls per night is mainly genetically determined. Some males call several hours per night and seldom change to satellite behaviour. Other males call little or not at all, even when they are kept in isolation. Many males can thus carry out both behavioural patterns but change with different frequencies from one behaviour to the other, based on their genetic disposition. They pursue different *strategies*. The behavioural alternatives of calling or silence within the strategy are called *tactics*.

Callers have greater reproductive success over a shorter time but do not live as long (fly predation). The low yearly reproductive success of satellite males is evened out by their long life expectancy. The life reproductive success is about the same for the representatives of the different strategies. Strategies that, during evolution, have not been replaced by other more successful strategies are called *evolutionarily stable*. If the number of parasitic flies increases, the number of callers decreases since they can be located and thus attacked by the flies. Males that call for shorter periods are less often affected, live longer and reproduce more successfully. However, if the number of flies decreases, the callers once more have an advantage.

In other species, behavioural polymorphism seems to be based on one genetically determined strategy that consists of several tactics. If all individuals possess the same genetic basis, the particular situation determines which behavioural alternative is carried out by the individual animal. The external and internal conditions decide how the animal reacts and thus this is called *conditioned strategy*.

Large male toads attract females in spring by using loud calls. Often, a smaller male sits quietly nearby.

Rules for decision-making by the toads seem to be:
— If no other males are present — call!
— If you are the larger male — call!
— If a larger male chases you away — call somewhere else.
— If the above are not possible, then become a satellite.

Since all males, during their life span, use all these tactics according to the same rules, this is a case of *conditional strategy*.

Parents invest in their offspring

The individual effort that one parent expends for one offspring and that thereby increases the young's chances to survive but prevents the parent from raising more offspring was defined as *parental investment* by the sociobiologist ROBERT L. TRIVERS. Maternal investment is distinguished from paternal investment. Mammalian mothers invest intensively during pregnancy and suckling and cannot have another young during that time. The more easily offspring can survive alone with increasing age, the less the mother has to invest in them. Instead, she strives to have more offspring. For this reason, weaning conflicts occur (see box).

Whereas maternal investment in most mammals is similar, the fathers in various species show basic differences. Their behaviour has only recently been the interest of researchers, especially in the area of primate research. The figure in the margin shows the average contact frequency per hour between fathers and the young of chimpanzee, gorilla and siamang. It also provides information about the contact quality. Fathers can be close to their young (contact distance), they can groom it and play with it, carry it or share food with it. A comparison shows that chimpanzee males look after their young the least, whereas gorilla fathers are active and siamang males are the most active fathers.

Only males that have a high probability of being the actual fathers of the young are assumed to provide paternal care. If fatherhood is questionable, then intensive investment in the young would merely incur high costs with no guarantee of passing on their genes.

The evolution of caring fathers is best understood in the light of monogamy. By taking over a great part of the brood care, the male can relieve the female so that it can recover faster from the last pregnancy, birth and brood-care phase. The help of the male reduces the distance between births so that the fitness of the male and the female (both values are identical in monogamy) is increased.

Females can lower the chance of fatherhood by copulating with several males. As a reaction to this selection pressure, a behavioural pattern has occurred that stops infidelity, called partner guarding. Males that guard their partners continually follow their females, especially during the fertile phase. Males, however, cannot always prevent the female mating with further males. In the sex organs of the female, competition occurs between the various sperm cells for the egg. This so-called sperm competition has led to the development of particularly large testicles during evolution.

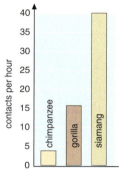

Paternal investment in various ape species

chimpanzee

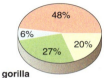
gorilla

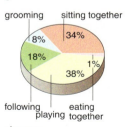
siamang

»info box«

Conflicts

Conflicts can occur within one individual if two contradicting behavioural patterns, e.g. attacking and fleeing, are triggered with the same intensity. Conflicts between individuals inevitably occur if the behaviour of one partner raises its own reproductive fitness but lowers that of the partner. As shown in the picture to the right, weaning conflicts occur regularly between mother and offspring. Conflicts between mating partners and between siblings similarly occur in many species, e.g. in some eagle species, this leads to the regular killing of siblings.

Weaning conflict in the Barbary Macaque

Behavioural Science

Reproductive strategies of the dunnock

Dunnocks (small singing birds) live in the dense undergrowth of forests, gardens and parks. The females build bowl-shaped nests in hedges or evergreen bushes.

Behaviour of the genders and mating

1 Breeding dunnock

In the spring, the females occupy a territory whose size depends on the supply of food. If a feeding place is installed, a smaller territory is defended.

The male territories cover the female territories. Males fight for territories that they can defend up to a size of 3000 m². The size here does not depend on the food supply.

If, in a monogamous couple, the female dies, the male often wanders away from his territory. If the male dies, the female usually stays in her territory. Neighbouring males whose territories reach into the female's territory try to expand their territory to cover all of the female's territory. If the area becomes so large that the male can no longer defend the whole area, then the males fight to establish a hierarchy over the whole space. The male highest in the hierarchy is called the α-male.

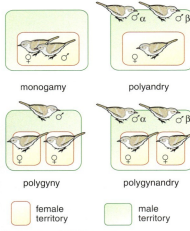

2 Possible social systems

On the basis of the various types of overlap of the male and female territories, four different combinations can be derived: *monogamy*, *polygyny*, *polyandry* and *polygynandry*. In polygynandry, the territories of two males that live together overlap with the territories of several females.

Task

① Use the information in the text and figures (figure 2 and 3) and summarize the conditions under which the various mating systems develop.

Figure 4 shows the relationship between the numbers of deaths and the number of snowy days in winter.

Figure 5 demonstrates the connection between the number of males and the number of polyandrous females.

Task

② Use the information and describe how the frequency of the mating systems can change from one year to another.

Competition between males

Males try to gain exclusive access to females by possessing territories or by hierarchy. Monogamous dunnock males conduct on average about 0.47 matings per hour. Males in polyandrous systems with exclusive access to females mate about 0.87 times. In the system in which both males copulate with the females, both males reach about 2.4 matings per hour. The testicles of dunnocks are about 64% larger than the size of mainly monogamous birds of same body size. In birds, the Fallopian tube terminates in the rectum so that the *cloaca* is also the transport path for the sex cells. As early as 1902, the ornithologist Schons

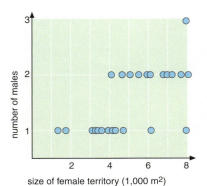

3 Size of territories

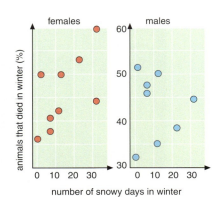

4 Winter mortality

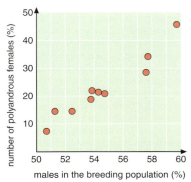

5 Breeding density

98 *Behavioural Science*

described a strange behaviour in the dunnock: "While the female stands with flapping wings and lifted tail presenting its cloaca, the male pecks at the cloaca a few times. The female then performs a pumping movement and secretes a droplet" (fig. 1). Copulation occurs immediately afterwards. Microscopic analyses have shown that this droplet contains sperm from a previous copulation.

1 Cloaca pecking

Freshly hatched young frequently disappeared or eggs were pecked in such a way that no young could hatch only in polyandrous and polygynandrous systems, in which the β-males (see p. 107) were unable to copulate.

The β-males were assumed to be responsible for this. They tried repeatedly to chase away the breeding female if they were not allowed to mate with it. In most cases, however, they were successfully hindered by the α-male higher in hierarchy.

Interestingly, such cases only occurred at the beginning of the breeding season when the females were likely to have further chances at breeding and not when the subsequent chances of breedings were unlikely. On losing her eggs, a females can re-lay eggs 1 to 2 weeks later at the latest. If they raise their offspring, they only start laying eggs again after about 6 weeks.

Task

(3) Use the described facts and work out the different mechanisms that are effective
a) before copulation
b) after copulation or
c) after egg laying
and that ensure reproductive advantages for the individual male.

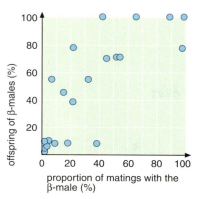

2 Breeding success of β-males

Paternal Investment

Male dunnocks can help their females either by incubating the eggs or by obtaining food to raise the nestlings. Help with incubating has been shown not to change the number of hatching young but the number of young successfully leaving the nest is significantly increased by assistance with feeding. This effect is the result of fewer nestlings starving to death. The help of the male was more effective depending on the number of young present.

During the mating season, the males guard the females. In monogamous couples (*guarding monogamy*), the male only drives those intruders away that could represent an additional mating opportunity for the female. The guarding is most intensive in polyandrous systems. Here the α-male follows the female all the time. However, he often loses sight of his female while chasing away the β-male. The female immediately hides in the dense bushes, eats peacefully and mates repeatedly with the β-male if she is found by him first. If she is found by the α-male, the game begins anew.

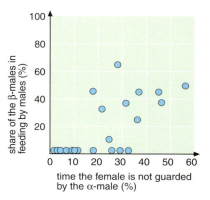

3 Contribution of brood caring by β-males

For sociobiological analysis, we need to know how successful β-male matings are. In order to examine this, a "genetic fingerprint" is made of all involved partners and the young in the nest. This makes it possible to assign fathers to the young. The result is shown in figure 2. Figure 3 shows the connection between the ability of the female to elude the α-male and the share of brood care taken over by the β-male.

Tasks

(4) Relate the results presented in figures 2 and 3.
(5) Explain the basis of the many conflicts that arise between the different partners by showing that the gain in fitness of one partner leads to a loss of fitness of the other. Evaluate the data in the table in relation to this.

Mating system	Number of young per year	
	per female	per male
polygyny	4.4	8.8
monogamy	5.9	5.9
polyandry (only α-male mates)	4.9	α: 4.9
polyandry (both α- and β-males mate and feed)	8.9	α: 4.9; β: 4.0
polygynandry	4.0	α: 5.6, β: 2.4

Behavioural Science **99**

1 Infanticide

Infanticide and reproductive success

"Eleven attacks the infant Fourfive several times and runs after the mother... The mother flees as fast as she can, whereas the older daughter gets in the way of the chaser. However, Eleven catches up with the female and pulls it to the ground. Moments later, six females attack the male. The mother grabs its baby from the resident male and continues fleeing. The canines of the male have ripped open the left thigh of the young, split the ischial tuberosity and cut open the lower part of the tail almost up to its root. Completely weakened, the infant dies during the evening of the next day." Such descriptions of aggressions of Indian monkeys, the *Hanuman Langur*, were interpreted as pathological excursions from normal behaviour. Sociobiologists today see it as an adaptation in the sense of genetic egoism because the killer increases its own reproductive fitness.

Hanuman Langurs live in groups of about 14 adult females and one male. The remaining males live in bachelor packs. The owner of the harem is regularly challenged by these maturing males. The continuous efforts at defending his group weaken the male to such an extent that it is replaced by a stronger male after about 26 months. The next male has about the same time to reproduce. At the time that the new male takes over the group, a few females are still pregnant, whereas other are suckling their young and will not ovulate during the suckling period. If they lose their young, the next ovulation occurs earlier. The behaviour of the new owner of the harem, i. e. his attempts to kill any young despite the resistance of the mothers (*infanticide*), has to be seen in this connection. He can now mate successfully with the females at an earlier time point. Extensive data have confirmed this hypothesis. The new owner of the harem kills mainly young that were not fathered by him. The mothers mate soon after with him. The interval between taking over the harem and having his own offspring is thus significantly reduced.

Observations have also led to the assumption that pregnant females are disturbed by the new male to such an extent that they abort (*feticide*).

The interpretation of infanticide and feticide as adapted behaviour can be tested experimentally. In an experiment, various numbers of eggs were removed from duck nests and observations were made as to whether the animals kept incubating the remaining eggs. In control tests, the eggs were replaced. In those cases that the animals leave the nest and stop incubating, two hypotheses can be proposed: on the one hand, a change at the nest and the loss of eggs could mean that the location of the nest and thus the breeding animals are in danger; on the other hand, the loss of the eggs could be so great that animals that continue to invest time and energy in the reduced brood raise fewer offspring than the animals that leave their nest and start a new brood with more eggs. The gain of time by deserting the original nest would thus lead to a gain in fitness instead of a reduction in fitness. None of the control nests without egg loss were deserted. The results of the set of experiments are demonstrated in the margin.

Tasks

1. Discuss which hypothesis is supported by the results of the experiments with the duck nests (see margin).
2. Describe, using the term fitness, the conflict between Langur males and their females.
3. Conclude from the perspective of sociobiology why the females mate with the new male.

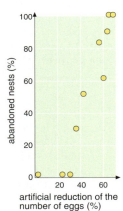

Resident male
owner of the harem

Manipulation of duck nests

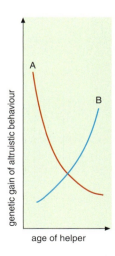

1 Florida Scrub Jay

Altruism
altruistic (selfless) behaviour, opposite of egoism

Hamilton's inequality
Altruistic behaviour is expected especially if the *costs* (C) for the altruist are smaller than the *benefit* (B) of the receiver and if the receiver is related as close as possible to the altruist. This is summarized in the so-called *Hamilton's inequality* ($C < r \cdot B$).

Coefficient of relatedness (r)
parents — children   0.50
true siblings   0.50
halft siblings   0.25
grandparents
— grandchildren   0.25

Altruistic behaviour

Even CHARLES DARWIN in the last century had problems finding an explanation for altruistic behaviour (*altruism*) as a result of evolution. His problem can be reduced to the question: How can inheritable units, which cause the permanent or temporary sacrifice of an organism's own reproduction, spread within a population?

In the *Florida Scrub Jay* that breeds in Florida in the undergrowth (scrub) of oaks, brood helpers (mainly young from previous broods) support their parents in raising their siblings. Thereby, the parents' chances of survival improve because they are assisted in rearing their young and the chances of the siblings surviving are better since they are protected better from predators.

Why do the helpers show this altruistic behaviour? The observations that helpers immediately occupy any territory that becomes free and give up the helping behaviour make it obvious that many of the males are hindered by the deficiency in territory in which to reproduce. If they cannot reproduce, helping their parents is a more reasonable alternative than doing nothing. From a genetic perspective, the males can increase their fitness with altruistic behaviour by supporting the reproductive fitness of relatives that also carry some copies of their own genes. Helpers increase the probability that younger siblings survive. Inheritable units that cause helping behaviour are thus indirectly represented more frequently in the next generation. If, with increasing age, being able to help their own parents becomes more unlikely (A) but being able to successfully fight for a territory becomes more likely (B), the helper changes if possible to breeding mode (see margin). The fitness of an individual is thus influenced not only by the number of its own offspring (*direct fitness*), but also by the number of offspring of relatives that survive as a result of additional help (*indirect fitness*). The sum of both values forms the overall fitness.

The closer the helper is related to the young, the higher is the contribution to its own fitness. The degree of relatedness is expressed by the coefficient of relatedness (r). It is defined as the probability with which a specific allele of an individual can be found in a relative. Since every diploid organism usually arises from the fusion of two haploid gametes, it has 50% ($r = 0.5$) of its alleles in common with each of its parents. The probability that a specific allele is passed on during meiosis is 0.5. If L is the number of generations in direct descent, the degree of relatedness of two individuals (the coefficient of relatedness) is calculated as $r = 0.5^L$ (see margin).

Animals that support their relatives have to be able to distinguish relatives from non-relatives. For this, they do not have to consider the degree of relatedness consciously but they should have mechanisms that control their behaviour as if they were indeed aware of the degree of relatedness. Animals can recognize the degree of relatedness, for example, by looks or smell or by components of a smell.

Thus, extending help to relatives means being able to pass on genes to the next generation by supporting relatives without reproducing by oneself. Thus, this should not really be called altruistic, since it is not.

Task

① Assume that an animal has the alternative to have its own offspring or to help its sister with raising her young. How many young must the sister additionally raise with this help in order genetically to compensate the sacrifice of the helper raising two of its own young?

Behavioural Science

Lifetime strategies

Organisms are selected to reproduce as successfully as possible during their lives, i.e. to produce as many surviving offspring as possible. In order to achieve this, individuals have to "make decisions" throughout their lives. Since the energy or matter gained by food intake is restricted, each organism has to "decide" for how long and for what they should invest their resources.

Growth or reproduction

1 Hermit crab

Hermit crabs hide their sensitive abdomen in a snail-shell. If the crab grows, it has to find a new home that is a slightly larger than the old one, but not too large. Well-fitting snail-shells are rare. Without a well-fitting home, they cannot grow further. Researchers provided hermit crabs with a limited supply of snail-shells and in a second experiment with an extensive supply. They examined the size of the crabs at reproduction and the size of the egg mass. They found, amongst other things, that crabs with an extensive supply reproduced at an older age (fig. 2).

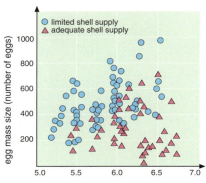

2 Reproduction of hermit crabs

Task

① Evaluate the text and figure 2 and interpret the data in the light of a lifetime strategy.

Reproduction and predator pressure

3 Guppies

Guppies, small freshwater fish, populate diverse environments in which they are chased by different predatory fish within their natural range. Closer examination showed that they were chased, in one environment (A), by a predator that was especially successful in hunting large adult guppies. In another environment (B), a second species of predatory fish hunted mainly younger smaller guppies. In both environments, these associations were measured and the data are presented in the following graphs. Here, the reproductive expense is the part of the biomass that females invest in reproduction. Females in environment (A) became sexually mature earlier. Breeding experiments in aquaria without predators showed that the different reproductive strategies were genetically determined.

Task

② Summarize the findings of figure 3 and relate them to the respective predatory strategies.

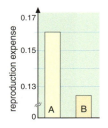 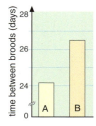

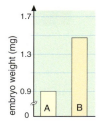

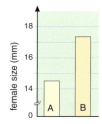 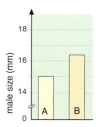

4 Reproductive strategies

Reproduction and survival rate

5 Great tit with young

Researchers examined various associations in great tits as shown in fig. 103.1 (a, b). In experiments, additional eggs were placed in the nests of the great tits, with the average clutch size being increased from 8 to 12 eggs. Later, the fledged young were re-captured (c).

Tasks

③ Summarize the results of the figures and relate the individual results.

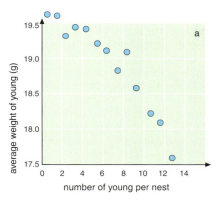

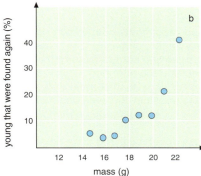

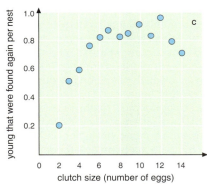

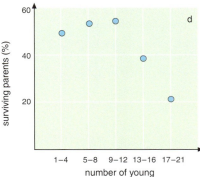

1 Reproduction of the great tit

④ Explain which factors determine optimal clutch size.
⑤ Discuss the influence of a good food supply on reproductive rate.

Parental investment

2 Herring Gull with young

Young animals have, statistically, an entire life of reproductive possibilities in front of them. The number of their potential offspring can be calculated. In older animals, part of their reproductive period has often already passed. The number of young that can be expected from such an animal is called the *residual reproductive value*. If the mortality of the parents rises, the residual reproductive value sinks. The animal must "decide" whether current intensive reproduction is more advantageous than later possibilities to reproduce. The older an animal gets, the less residual reproductive value it has to lose.

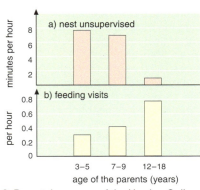

3 Parental expense of the Herring Gull

Figure 3 shows the parental investment of Herring Gulls. Part a shows the time during which the parents leave the nest unsupervised and part b shows the frequency of feeding visits per hour.

Task

⑥ Summarize the facts mentioned in the text and relate them to the figures.

Lifetime strategy and gender

4 A pyramid of the common slipper shell

The **common slipper shell**, which lives along sea coasts, forms small pyramids out of several animals of different sizes sitting on top of each other. The lower larger animals are always female, whereas the smaller animals on top are always male. Adolescents change their sex. Such sex changes also occur in various fish species. In some of these species, the larger animals become female, whereas in others, the larger individuals become male.

This change is explained as follows: if the reproductive success of one sex with increasing age or body size exceeds the reproductive success of an animal of the same age or size but of the opposite sex, the animal should change to the more successful gender. Thus, a smaller animal can produce many sperm cells and, with them, fertilize more egg cells than it could produce itself (an individual has to expend more energy and material producing eggs than sperm cells). With increasing body size and high sperm competition of the males, the number of egg cells that a large fish as a female can produce might exceed the number of successful sperm that would be present if the same fish was male. This means that the "decision" of a male to become a female also depends on the number of male competitors that are already present.

Tasks

⑦ List the conditions under which adult common slipper shells become female or male respectively.
⑧ The more common slipper shells that form a pyramid, the earlier the lower animals become females. Please explain.

Behavioural Science

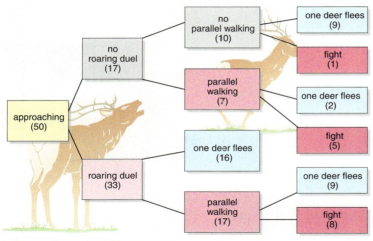

1 Observed course of 50 contests

Fighting strategies of Red Deer

As in many other animal species, two different fighting situations can arise for red deer: defence against predators and within interspecies competition for resources or partners. Deer flee from predators such as wolves or defend themselves by using their hooves. In interspecies fights, they use their hooves if their antlers are not yet usable. The antlers are only used in autumn during the rut.

For most of the year, male and female animals roam in separate herds within their territory. They gather in traditional rut areas only during the rut. Male deer try to collect a group of females, a harem, around them. This can include up to 20 females. Offspring are fathered with the owner of the harem, the dominant male, as soon as the female deer signal preparedness for conception.

Only a few males can found and maintain a harem, because competition is fierce. Rivals try repeatedly to lure one or even all the female deer into their own harem in order to increase its size. Fights between rivals are thus always possible and can be observed frequently. Deer show a graded fighting approach, a strategy, that allows the progression of a fight (*escalation strategy*).

In general, a deep bellow is heard originating from the male deer during the rut. This is called roaring. The tone pitch and intensity depend on the body size and condition of the animal. In the beginning, the roaring is general and not directed to a particular rival.

However, if a rival approaches, an increasing interplay of roaring, i.e. a roaring duel, often begins. The individuals can apparently assess each other even at this stage so that some contests end here. Otherwise, the rivals start walking in parallel at a distance of 5 to 10 m during which the opponent is again assessed. Termination of the fight is still possible at this stage. However, if this does not happen, the opponents lower their antlers and run at each other with great force. Each tries to push the opponent away or even to hurt him. Most animals are apparently skilful enough to avoid harm but, nevertheless, 20 — 30 % of all deer suffer permanent damage during their lifetime because of these fights. A fight is terminated if one opponent is severely injured or flees.

In principle, deer fights contain two sub-strategies. On the one hand, one tries to impress, push and compare strength without hurting the opponent. This is called *ritualized fighting*. On the other hand, severe injuries are possible and even death can occur. This is called *injurious fighting*.

The stages observed in red deer, namely to start with ritualized fighting and then proceed to injurious fighting only if necessary, seem to be generally advantageous. Only the least necessary effort is undertaken. However, this overall strategy is not found in all animal species. In some, only ritualized fight occurs, whereas others indulge mainly in injurious fighting. The question as to which conditions favour the one or the other behavioural pattern is examined by conceptual methods (see page 105). Here, attempts have been made to describe the fighting strategies and their success by using mathematical models. The conditions that make a comment fight or an injurious fight more advantageous in the model can be used to explain real observable behaviour in nature or to form the basis of further investigations.

»info box«

Computer simulation of fight strategies

Model with two strategies

Animals fight with other individuals of their species for a resource. In the case described, this is a mating partner with whom offspring is produced. In the simulation, we wish to know under which conditions one of the fight strategies, namely ritualized fight (r-fight) or injurious fight (i-fight), is more successful.

The model is constructed as follows: in one population, individuals carry out r-fight or i-fight.

Ritualized fighters (r): only r-fight, flees in i-fight with opponent without injuries.
Injuring fighters (i): only i-fight.

In the model, the population is composed of r-fighters at the start (e.g. 99%) to which i-fighters are added anew because of mutations. All possible fight arrangements are simulated and their results are determined. These could be defeat or victory, plus additional injuries when defeated or when having an i-fight, that are evaluated as follows:

defeat: always 0 points,
victory: +50 points,
injuries: −100 points

	evaluated fighter	
	r	i
r	25	50
i	0	−25

For the same probability of two results, the average is counted, e.g. i-fighter against i-fighter:
$(50 + (-100)) : 2 = -25$

The exact values for the points are allocated randomly. This is not of importance because, later, the important results are the differences that arise from the evaluation with points.

When calculating the values for a new generation, the points reached are calculated from all possible fighter combinations for each fighter type. The new generation contains both fighter types according to the ratio of the points reached. During the course of a generation, an evolution process occurs that changes the frequencies of the strategies.

The comparison of different simulations shows the following. An increase in victory points or a decrease in injury points favour i-fight. This shows a relative weighting of both factors: the greater the potential gain (victory points), the more successful it is to take a higher risk (injury points). Often certain parts of both strategies remain; the fighter types are both equally common according to the balance ratio. This can also mean that all individual use both strategies as a mixture. Here, the model with two strategies is shown to be constructed in too simple a manner and needs an addition that contains combinations of both strategies as a tactic.

Model with five strategies

The model with five strategies contains, in addition to the two known strategies, three strategies that also acknowledge a change of tactics:

Explorer (E): starts with r-fight, changes to i-fight if the opponent shows r-fight, but flees in i-fight.
Revenger (R): starts with r-fight, changes to i-fight if the opponent shows i-fight.
Intimidator (I): starts with i-fight, flees from i-fight of the opponent.

In this model, the revenge strategy is superior. Sometimes, a few pure comment fighters occur. However, this is impossible to distinguish later on because only r-fights will occur between revengers and ritualized fighters. All individuals thus seem to be ritualized fighters. Therefore, even this model has its limitations, since there is no intention of increasing aggression, even if the gain makes it seem worthwhile.

Tasks

① Make a table with all fighter arrangements and fill in the fight results.
② Examine to what extent the fight strategy models deliver explanations for the behaviour of red deer.

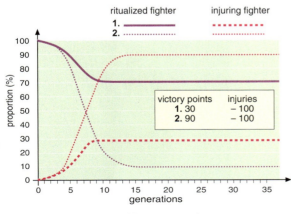

Result of the simulation with two strategies

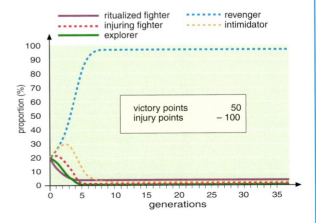

»info box«

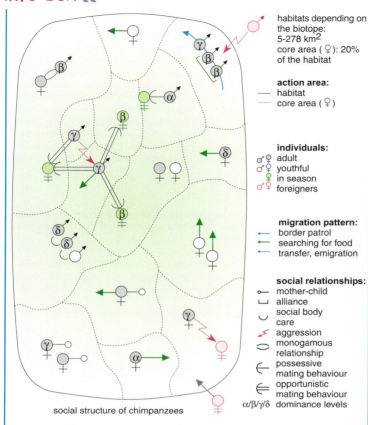

social structure of chimpanzees

habitats depending on the biotope:
5-278 km²
core area (♀): 20% of the habitat

action area:
— habitat
···· core area (♀)

individuals:
♂♀ adult
♂♀ youthful
♀ in season
♂♀ foreigners

migration pattern:
← border patrol
← searching for food
← transfer, emigration

social relationships:
⊸ mother-child
⊔ alliance
⌣ social body care
⚡ aggression
⊖ monogamous relationship
⇐ possessive mating behaviour
⇐ opportunistic mating behaviour
α/β/γ/δ dominance levels

Chimpanzee society

Chimpanzees live in groups of up to 50 animals and preferably eat fruit. Depending on the habitat, the density of ripe fruit trees varies greatly. Thus, chimpanzee society is distributed into male groups, mother-child groups and groups of mixed sexes depending on the food supply and the number of females in season. The gathering and dispersal of subgroups are referred to as a *gathering-dispersal society*. One advantage of gathering is the possibility to hunt cooperatively in groups; this increases the probability of catching small animals of prey such as small monkeys, bush pigs or young antelopes. Hunting is mainly carried out by males, who later share their prey with the females.

Males defend the group's territory together. Alone, they cannot maintain their goal to have access to as many females as possible. In this way, the male group together secures the most important resource (females ready for mating) since females can roam a broad area when searching for food. Foreign females are accepted into the community, whereas males are not. Females born in the group leave the group before sexual maturity and males remain in the group. They are related and know each other well. Male chimpanzees form alliances.

Female chimpanzees copulate with as many males of the group as possible, which leads to the result that all males could be the father. This is a possible reason for none of the males acting aggressively against the newborns. In West Africa, females secretly change groups during their fertile phase so that about 50 % of all young are fathered by males from neighbouring groups.

Aggression and social hierarchy

Originally, chimpanzees were considered as extremely peaceful animals but research over the last few decades has shown the exact opposite. Chimpanzees are extremely aggressive and can be a deadly thread for other chimpanzees. Two forms of aggression can be distinguished based on intensity: fights against foreigners to the group and conflicts between members of the group. Fights against members of neighbouring groups are almost exclusively between males, take longer than fights within the group and are also particularly brutal. About 30 % of all chimpanzee males die in such inter-group conflicts. In fights against neighbouring males, chimpanzees behave more as if they were hunting for prey. The chimpanzee researcher JANE GOODALL writes: "I assume that if they were given guns and explained how they work them, they would use them to shoot their neighbours."

Moreover, 30 % of the fights within a group lead to injuries. Whereas females fight mainly about access to food or to defend their young, males fight especially often for access to females that are ready for mating and with respect to their place in the hierarchy. If females high in the hierarchy help their young in fights, their rank is passed on to their offspring. Subsequently, they have especially successful sons that will generate many grandchildren.

1 Display behaviour in chimpanzees

Since, according to BATEMAN, differences in fitness can differ more often in males than in females, there is more at risk in conflicts for males. This is why they fight more intensively than females.

Hierarchies are present in many closed individualistic societies. They are sometimes linear and animals higher in the hierarchy dominate the animals lower in hierarchy. Based on the Greek alphabet, the animal highest in the hierarchy is called the α-animal, followed by the β-animal, and then the γ-animal. The Ω-animal is at the lower end of the hierarchy. Other hierarchies consist of groups of animals of the same rank and also triangular relationships with partly inverted hierarchical positions between individual animals.

Hierarchies lead to privileges regarding access to resources and thus to differences in the success of fathering or rearing offspring; this results in fitness differences. Hierarchical fights occur or are avoided depending on the extent of the gain or loss in the conflict. A fight is worth it if the benefit of the resources (B_R) multiplied by the probability to win the fight (P_W) exceeds the expected costs of the fight (C_F), i.e. $B_R \cdot P_W > C_F$.

Whereas the hierarchy is usually clearly defined between males, this is usually missing for females. Males high in the hierarchy have advantages in mating with sexually mature females, whereas females high in the hierarchy have better access to food. Females ready for mating copulate with many males one after the other, whereas the male highest in hierarchy mates many times at around the time of ovulation so that his sperm most probably fertilize the egg cell. Sperm of different males compete for the egg cell. During evolution, this has lead to the formation of especially large testicles. Weak males secure offspring by leaving the group with one female before she shows a readiness to mate.

Within the hierarchy, losers experience disadvantages, although these are usually compensated by the advantage of living in a group. On the one hand, animals lower in hierarchy use the protection of the group against enemies. On the other hand, losers have fewer expenses and risks at times in which success in a fight is improbable. They wait until they can later take over the lead role or change group.

Task

① Evaluate, based on risk analysis, why males generate hierarchies more often than females.

»info box«

Reconciliation

The primate researcher FRANS DE WAAL observed that, in chimpanzees, aggressive encounters are followed, in 40 % of the cases, with the opponents coming into contact within about half an hour in order to reconcile with each other.

In the simplest case, a hand is extended towards the opponent with the palm facing upwards. If males are not able to communicate again after a fight, a female can mediate between the two. She approaches one male and carries out social fur care (grooming) and, at the same time, encourages the opponent by eye contact to follow her example. When the second male sits next to the female, both start to grooming the first male. The female then retreats and the two males groom each other.

Whereas reconciliation behaviour in chimpanzees usually originates from the inferior animal, in bonobos, the superior animal of two previous opponents takes the initiative. This chimpanzee species often reduces social tension by sexual contacts. These examples demonstrate that reconciliation and the avoidance of conflicts is genetically predetermined in our close relatives.

Behavioural Science

Complete truthfulness
Interests are shared completely by the transmitter and receiver.
Example: the dance of honey bees, insects having warning colours.

Limited truthfulness
A high degree of shared interests limits deception. The truthful core is large.
Examples: spouses, parents and children, siblings.

Signals and communication

Merely by existing, every animal sends signals out into the environment. It reflects light so that possibly colours and movements are visible and it releases scents or sound.

If an animal (the *transmitter*) sends a signal that is received, processed and reacted to by another animal (the *receiver*), this is usually referred to as communication (fig. 1). Such communication systems can occur not only between different species, for example, between predator and prey, but also within one species.

For each signal, the fitness advantages or disadvantages can be estimated for the transmitter and the receiver. Three combinations, as demonstrated in figure 109.1, are then distinguished:

2 Red Uakari

Early on, deception and cheating were realized to occur during communication between different species; this is of course not consciously done but is mostly highly effective. Because classic ethology assumed that animals do everything to maintain their species, communication within a species was expected mainly to be made up of cooperative signalling and cheating was not taken into account. The concept that individuals can act egoistically eventually made it clear that deception also often occurs during communication within one species. Genetic fingerprinting has surprisingly shown that, in some bird nests, eggs are present that come from another female (*brood parasitism within a species*). The degree of truthfulness lies somewhere in between the two extremes of complete truthfulness and complete deception, depending on the environment and requirements of the transmitter and receiver.

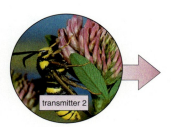

1 Transmitter-receiver model

1. If a poisonous animal has a warning coloration and is therefore not eaten or if bees inform their hive-mates about a food source, this is *cooperative signalling*.
2. *Accidental signalling* occurs if a birds looks for a female by singing but, at the same time, reveals its position to a bird of prey.
3. *Deceptive signalling* is carried out, for example, by a harmless insect that mimics a poisonous exemplar.

Deceptive signalling is increasingly probable,
— the easier it is to falsify a signal,
— the fewer interests are shared by the transmitter and receiver,
— the lower the degree of relatedness between the transmitter and receiver,
— the less often two partners meet, because in this case, mutual advantage considerations become less probable.

Limited deception
Deception is only limited if other species members display distrust. The truthful core is small. Example: signals between rivals, especially if they are not related.

Complete deception
No shared interests between transmitter and receiver. Example: predator and prey.

	fitness effect for	
	transmitter	receiver
cooperative signalling	⊕	⊕
accidental signalling	⊖	⊕
deceptive signalling	⊕	⊖

1 Communication and fitness effect

Avoiding deception

Deception can be avoided in two ways by the receiving side. First, the receiver can react only to signals that are difficult to falsify or are not falsifiable. Many females of the animal kingdom (see info box) evaluate the survivability of males that are potential mating partners by features that cannot be falsified by younger or less healthy rivals, e. g. the possession of decorative feathers that hinder flying (*handicap principle*). The songs of blue tit males become more and more complicated with increasing age and represent a non-falsifiable sign for good survivability. The face colour of the South American red uakari is only bright red in males that have a functional immune system (figure 108.2). Sick males are paler. This also applies to the red belly of male sticklebacks. Females can thus reliably choose a healthy partner by his colour.

Second, the receiver can avoid deception by being distrustful. The evolution of the human ability to talk certainly brought great selective advantages but also created the problem that spoken information can easily be falsified. This probably led to us placing trust in those individuals who are most likely to be honest, e. g. relatives and well-known acquaintances whom we meet daily. Many researchers believe that the development of human speech was accompanied by the development of means to recognize members of our own group e. g. by dialects and clan symbols.

Task

① Describe how cosmetics, fashion and cosmetic surgery could be exploited as signal deception.

»info box«

Ritualisation

The *common pheasant* scratches about on the floor during courtship, moving back and attracting the hen by pecking food from the ground. The hen comes closer. The *Chinese monal* develops a fan-shaped tail during the courtship period. It bows in front of the hen, slightly spreads its tail and pecks with its beak at the ground. The hen comes closer and looks for food in vain. The male now spreads its tail and wings to the maximal extend. Its tail feathers move in a slow rhythm back and forth. The *peacock pheasant* scratches on the ground and bows while raising its wings and spreading its tail. If a female approaches, the male offers food to her. The behavioural patterns during courtship are derived from "food attraction". Hens get their chicks to come towards food by scratching the ground, pecking and making attracting calls. The comparison of the closely related pheasant species shows that the behavioural patterns became removed from their original functional sphere and were newly incorporated into reproductive behaviour. Hereby, gestures evolved that are carried out extremely slowly and that are emphasized by flashy colours and frequently also by rhythmical movements. During evolution, "food attraction" took on a new meaning. It acquired signal functions (*ritualisation*) and now serves in communication within one species.

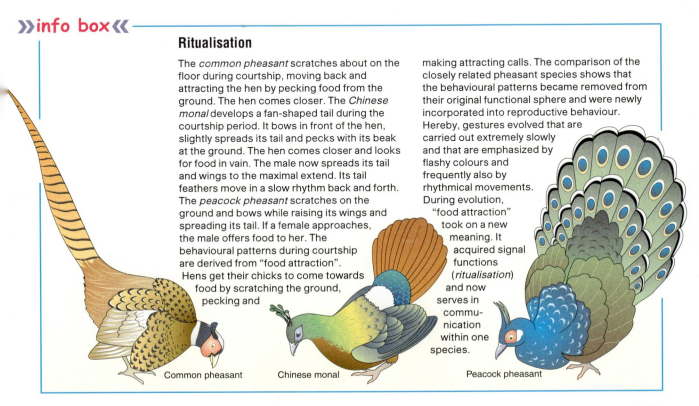

Common pheasant   Chinese monal   Peacock pheasant

Behavioural Science

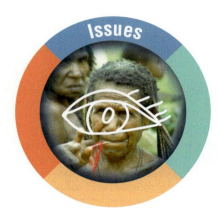

Cultural diversity and universalism

Cultural diversity

The ability to learn enables humans to adapt their behaviour to rapid changes or local characteristics in the environment and to acquire this behaviour from the preceding generation. Habitats such as cold steppes, rainforests or savannahs require different tools and nutritional strategies. They also encourage the setting up of various social systems and possibly also of moral standards and religious beliefs.

Traditional ideas

Traditional anthropology saw the human brain as an all-purpose computer that is empty at birth but that is later programmed by external sources. Based on this idea, human behaviour was considered to be entirely a product of society. The anthropologist BROWN summarized the accepted theories as follows: "Culture proceeds randomly and can take on countless variations within one society.

Human behaviour is basically determined by culture and not by biology or genetics." Anthropologists have argued from this standpoint from the end of the 19th century until today, with a few excpetions. Common biologically determined behavioural patterns were thus considered neither necessary nor possible.

The new approach

More recent investigations have shown that the perception of an empty all-purpose computer-brain that is programmed from the outside is wrong. TOOBY and COSMIDES have established that the human brain contains many "modules" that are responsible for different fields of problem-solving and that control behavioural patterns. These must be the same for everyone (universalisms) because they evolved during the about 2 million years in which our ancestors lived in hunter-gatherer cultures. Within the few thousand years in which modern cultures such as farming have existed, no evolutionary changes could have affected the human brain.

The conclusion consequently is that all people worldwide must have a common "basic design".

Acquiring culture

Until recently, culture as acted out by parental example was thought to be passively acquired by children (*tradition*). Subsequently, we have realized that this acquisition by the children is both a highly active and a selective process. Children pick out "cultural units" from what surrounds them. Thus, this is referred to today rather as "reconstructed" or "adopted" culture. Every child is born into a group with behavioural rules and moral beliefs. It observes the behaviour of the group members, learns the dominating rules of the group and listens to the opinions of the others

in order to be able subsequently to predict possible reactions and behavioural patterns. Only in this way is communication possible. Hereby, it is especially easy to learn from relatives, friends and successful people. If the beliefs of the communication partners diverge, then communication becomes difficult. This is why we experience "culture shock" when visiting foreign countries. The acquisition of moral concepts and ideals seems to have a character similar to imprinting so that, once learned, concepts can only be changed with difficulty or not at all.

Universalisms

Features that all people have in common are called universalisms. These include:
a) a world view with themselves at the centre; people "humanise" animals and objects, think in causal chains and not in networks and can imagine transcendent beings.
b) group living in closed structured groups; people in groups imitate each other and understand each other, prefer their relatives, are able to detect social cheats, create dialects, clan and status symbols, see themselves as the "one and only people" and distrust foreigners (ethnocentrism), develop common concepts of standards and love gossip.
c) sexuality; people have developed an incest taboo, have common criteria for choosing partners and show types of jealousy that differ gender-specifically.

Think about what impact puberty could have on culture acquisition.

For the Inuit who eat mainly marine animals, their most important goddess Sedna rules the animals of the sea and lives fittingly at the bottom of the sea.

Gather information about the differences and similarities between the various religions.

Consider why tools with the same function show greater conformity than beliefs in gods.

Searching for partners

Studies in various cultures worldwide have shown that the criteria used by men and by women differ with regard to their choice of partner. However, the criteria within populations of men and of women mainly conform.

Pleasant relaxed male, studious, athletic, slim, 48 years old, 177 cm, nature-lover, sensitive, wants to meet slim, younger, good-looking, well-educated woman with high standards, with view to a happy life together. code: 684150667A

Female university graduate, 52 years old, 162 cm, juvenile type, blond, 50 kg, normal figure, would like permanent relationship with warm-hearted relaxed partner. Preferably with picture. code: 157468066A

Analyse the marriage advertisements and determine which properties men and women praise with regard to themselves and what they expect from potential partners (e.g. age, income). This is also possible via the internet in foreign countries.

Tools and gods

Approximately 20,000 years ago, all cultures were probably similar. People used stone tools that differed worldwide only in their detail. All differences that we perceive today have subsequently been developed over time by the people in their local populations. Although their tools, depending on their function, are often similar, decoration, body jewellery or their beliefs in gods differ greatly.

Nevertheless, conformities can often be found. Hunter cultures that orient by the stars on their wanderings often localize their gods in the sky, whereas the gods of farmers are often female and associated with the earth.

Choice of habitat

A habitat that suited our ancestors had to fulfil various criteria: it had to provide food and water, to be manageable and to give protection. Trees were certainly important for protection but only if they were easy to climb and acted as a screen. Since the savannah has been the habitat of our ancestors for the longest time, we might expect that the features of this landscape would also be preferred by people today.

Collect pictures of various landscapes (desert, steppe, savannah — with and without water — or tropical rainforest) and test, in interviews, the habitat preferences of fellow pupils.

Test, by means of various pictures, which features a tree must have in order to encourage people to have a picnic close by.

Behavioural Science

Basic concepts

Structures and functions

Organisms are characterized by the fact that most of their structures, and the processes they carry out, help them to function in some specific way (see fig. 1). When we see a feather or a flower, we should ask ourselves about their biological functions. This does not apply to salt crystals or stones. In nature, structure and function are only connected in living organisms. This can be explained by the evolution of the organisms and can be identified by using different models.

The lock-and-key principle

All organisms contain numerous molecules with specific structures. These molecules can interact with other molecules to which they fit sterically. For example, this lock-and-key model allows enzymes to interact with a substrate or transmitter molecules to bind to their receptors. Whether the binding partners can fit properly into one another is determined by their structure, their shape and the distribution of their electric charges. These factors form the basis of each specific interaction.

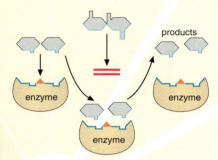

The lock-and-key principle

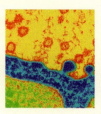

Presynaptic membrane, synaptic cleft and postsynaptic membrane

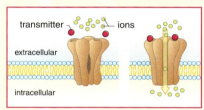

2 What is the function of the transmitter? Where do you think it comes from?

1 How do the hairy peds help a fly to walk over surfaces? Do they help a fly to "stick" to the ceiling? Explain!

The building block principle

Macromolecules are made up of identical or similar subunits; simple sugars, amino acids and nucleotides determine the highly specific structure of polysaccharides, proteins and nucleic acids. Even the smallest changes in the subunit may lead to a loss or change in function. On the other hand, when similar subunits are slightly modified, they might be able to carry out completely different functions. Examples of such combinations are: the porphyrin system found in haem, cytochrome and chlorophyll; ATP, either as an energy carrier or when it forms AMP, which is a part of DNA and RNA.

The building block principle

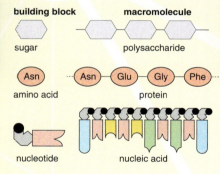

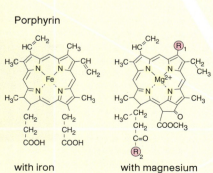

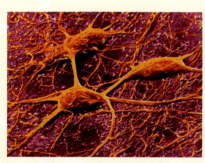

The modification principle

All organisms are made up of cells. Cells are the smallest units that live and reproduce, and they have similar basic characteristics. Differentiation, however, results in structural differences between cells and these differences indicate their various functions. For example, when we examine the structure of a nerve cell, we can clearly see how it functions (see figure).

The modification principle

● Think about connections. ● See connections. ● Develop connections.

112 Basic Concepts

The antagonism principle

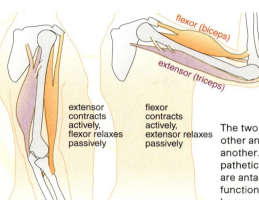

extensor contracts actively, flexor relaxes passively

flexor contracts actively, extensor relaxes passively

The antagonism principle

We call flexor and extensor muscles, and muscles and expandable tendons antagonists. When the antagonists interact, movement is possible. The constriction and dilation of the iris muscle causes a narrowing or enlargement of the pupil, and we also call this an antagonistic movement. Contraction is an active process, whereas relaxation is passive.

The two processes work against each other and are coordinated with one another. The sympathetic and parasympathetic neurons in the nervous system are antagonists which regulate organ functions. In metabolic processes two hormones, such as insulin and glucagon, can work antagonistically.

The countercurrent principle

A countercurrent mechanism is used in the recycling and concentration of substances and heat. The mechanism is based on diffusion and heat conduction and is maintained when the concentration gradient between the transporting media is as steep as possible. One example of this mechanism is when cold blood from the periphery flows into a venous capillary system located close to arteries. The arterial blood is cooled and the venous blood is warmed up. This mechanism is used by a goose when it stands on ice: its feet can almost reach freezing point but it only loses little heat energy.

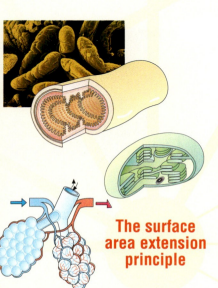

The surface area extension principle

The inner membranes of mitochondria and chloroplasts contain enzymes that are arranged in a fixed steric order. There are many folds in the membranes of both organelles. The folds significantly increase the surface of the membranes, thus increasing the number of reactions that occur and hence the metabolic rate. The principle also applies to organs: the surface area per volume unit that is available for absorption is increased in the lungs by numerous tiny sac-like structures and in the intestine by abundant folds, villi and inversions.

Tasks

① Use examples to explain how an ordered state in a cell's structure also means an ordered state in its function. A disordered state in its structure, however, often means a change or a loss in function.

② Antagonistic effects are found in metabolic, growth and developmental processes. Explain how antagonism works as compared to synergism (harmonic cooperation), giving an example for each of the three processes.

③ Dogs, sheep and other mammals have a so-called *Rete mirabile* (lat.: wonderful net). This is made up of hundreds of arteries that run parallel to one another and are derived from the large head arteries. The Rete mirabile is surrounded by a bubble-shaped vein containing blood from the mucosa in the mouth and nose. How does this help a dog to chase its prey, sometimes causing the exhaustion and death?

• Think about connections.   • See connections.   • Develop connections.

Basic Concepts

Basic concepts

Regulation and control

Organisms and biocenoses keep many state variables within close limits, even when internal or external factors change for a short time (homeostasis). Through regulation systems, these state variables remain within a functionally adequate range. According to technical cybernetics, the aspired value is defined as the *set value*.
For certain procedures it is possible to adjust this value to changes that are happening. While a state can be maintained through regulation, control mechanisms can modify the direction or intensity of processes.

Negative feedback is a characteristic feature of regulation systems. Models of *control loop* systems in the form of block diagrams can show this by distinguishing between different components of a control loop. They are used to explain physiological processes and are usually complex. Arrow diagrams are easier to understand and are often used in population ecology.

Pupil reflex

The pupil reflex is an example for a proportional control system: with increasing light exposure the pupil becomes smaller, while less light exposure enlarges the pupil according to the current deviation from the set value.

Contrary to the simplest form of a control loop, in this system there is a superior control centre, which allows the adjustment of the set value. In simplified terms this can be compared with a water level control system.

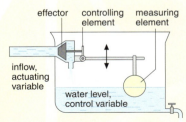
automatic water level regulation

Pupil reflex

Blood sugar level

The level of blood sugar is regulated by several interlocking control loops (networked systems). The concentration of blood sugar is constantly measured and maintained within a range of 60 to 120 mg glucose per 100 ml of blood. Two hormones produced in the pancreas (insulin, glucagon) are involved in this regulation system. Their activity is influenced by other organs, nerve signals and other hormones. Factors like stress, muscle activity or times of rest act as disturbances.

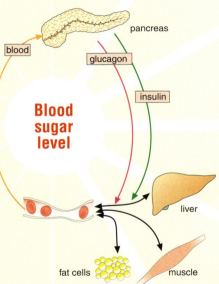

Body temperature

Birds and mammals are able to maintain an almost constant body temperature. Two very different effectors are involved in this regulation system: muscle contractions are used to actively release heat; sweating and panting is used to cool down.
For poikilothermic animals, behavioural changes can sometimes act as effectors: reptiles living in high altitudes or in the steppe regulate their body temperature through sunbathing during the day and by retreating to protected caves during the night.

Body temperature

- Think about connections
- See connections
- Develop connections

Examples from technology

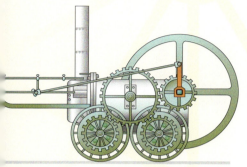

Examples from technology

Control loop models mirror models used in technology: thermostats of heaters and refrigerators or the controls of an iron keep *preset* temperatures constant. However, since the construction of the first steam engine, other regulatory principles have also played a role in industry and technology: in 1788 JAMES WATT (1736 – 1819) developed a *flyweight governor* that ensured a constant output performance of steam-powered machines. Control systems for coolants in cars are e.g. linked with pipe systems that have the maximum area for heat exchange possible.

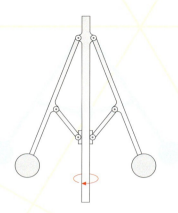

Population size

The population sizes of different species of a biocenosis are interdependent. Feedback systems deliver only temporarily stable conditions; they can change quickly to reflect new environmental conditions. A defined set value does not exist.

An increase in the number of individuals can for example intensify intraspecific competition; it can lead to an oversupply of resources for predators or parasites and to an increased danger of spreading of infectious diseases.

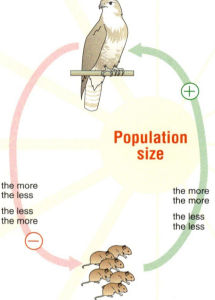

Population size

the more the less
the less the more

the more the more
the less the less

Tasks

① Clarify the differences between regulation and control systems using the examples of motorised toy boats with and without remote control.

② Contrary to the pupil reflex, the regulation of blood pressure is an example of 'temporal integral regulation': despite the influence of permanent disturbance variables, the regulation process achieves and maintains a must value. Explain this with regard to blood loss due to injury.

③ The principle of negative feedback reduces or completely compensates for deviations from a set value through counteractions. Other models are mechanical equilibria, overflow systems, chemical buffering or 'feed-forward systems'. Find information about these models and comment on differences.

④ Positive feedback is self-intensifying. Find an example for this statement in the field of both biology and sports. Explain how each example works.

⑤ Muscles receive signals to contract both via α motor neurons and via β fibres, which lead to muscle spindles. Prepare a short presentation about the mutual influences between muscles and muscle spindles with regard to contraction and elongation.

Genetic and enzymatic activity

Repression of end products and substrate induction are examples for a negative gene regulation as they are controlled by the feedback of repressor proteins. Positive gene regulation is arranged through the attachment of activating proteins on a control region. In both processes allosteric proteins play a role. They can modify their spatial structure through the attachment of effectors (conformational change).

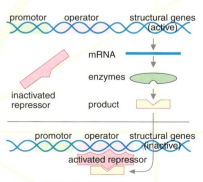

Genetic and enzymatic activity

Think about connections   • See connections   • Develop connections

Basic Concepts   115

Basic concepts

Information and communication

Organisms receive, save and process information and communicate with each other. Precondition for this are a common language and suitable receiver, storage and sending mechanisms.

The word "information" has various uses in everyday language: objects such as a CD *contain* information, and organisms *receive* information. In biology, we define information as a message that is made up of a spatial or chronological sequence of signals that causes a certain effect. Communication is a mutual transfer of information that is adapted to the communicating partners. It can occur between organisms but also within an organism and inside a cell.

Sender and receiver

In a communication chain, the sender transmits an encoded signal that not only must arrive at, but must also be decoded by the receiver. This means that signals and processing mechanisms must be compatible. One example is *pheromones*. These substances serve in the communication between individuals of one species. For example, *bombicol* — the sexual attractant of the silkworm moth — is only produced by the females. The appropriate receptors occur in the large antennae of the male moth and react even to minute concentrations of this substance.

Sender and receiver

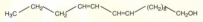

structural formula of bombicol

pheromone receptors on a sensory hair

Encoding / Decoding

A code involves specific instructions for converting information from one language to another. Examples are the Morse code, the encoding of texts in bits and bytes, and the telephone. With the genetic code, information in "the language of nucleic acids" is translated into "the language of amino acids".

By encoding information, transfer can be made safer. Nerve cells send messages encoded by frequency and process their modulation by amplitude.

Encoding / Decoding

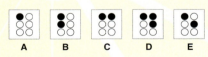

Redundancy

The term "redundancy" is derived from information theory (Latin: *redundantia* = surplus) and describes the informative surplus of a message. In communication engineering, high redundancy protects a message from disturbances during transmission.

In genetics and evolution, redundancy is the multiple existence of similar signal structures. This ensures stability and reliability. Parsimoniously (according to economical principles), a minimal number of participating factors is used. Only five elements make up DNA, four bases encode the entire variety of all species, and about 20 amino acids form all proteins.

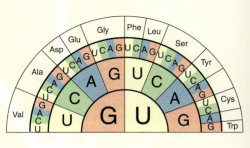

Redundancy

● Think about connections ● See connections ● Develop connections

Flow of information within a cell

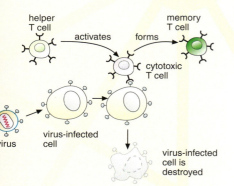

Flow of information within a cell

Within a cell, continuous information transfer occurs from organelle to organelle via various chemical substances. Adjacent cells within a tissue can also communicate with each other. Special cell-cell contacts make the exchange of small molecules possible. In the immune system, various cells communicate with each other either by signalling substances or by direct cell contact.

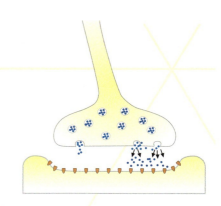

Nervous and hormone systems

The nervous system is used for fast and precise communication. Not only electronic signals, but also chemical signals are employed. The interaction of afferent and efferent routes influences the flow of information by means of feedback mechanisms. In addition to the fast effective transport of information by the nervous system, hormones are used in the blood stream for long-term communication. They have an effect on the corresponding receptors of their target organs.

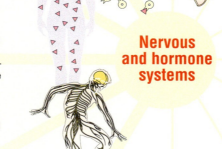

Nervous and hormone systems

Communication

The term "language" applies primarily only to verbal communication between people. In the broader sense, it also includes symbolic languages (sign language) or the language of animals (e.g. the dance of the honey bee or body language). In social organisations, optical and acoustic signals for communication are important for social cohesion or for collective activities. Plants also communicate with each other or with animals, for example, by releasing warning substances when being attacked by pests.

Communication

Tasks

1. Think about one example each from behavioural biology, neurobiology, metabolic biology and genetics showing how the sender (transmitter) and receiver are adapted to each other.
2. Short waves and medium waves in broadcasting have a significantly worse quality than very high frequency stations. Find an explanation.
3. Present different encoding mechanisms for information transfer in the nervous system.
4. Plants save information regarding light exposure conditions in the form of phytochrome molecules. Give details of the function of this molecule and explain how information is processed by it.
5. Books, movies, a CD or DVD are examples for technical storage media. Compare these data carriers to biological information carriers regarding the kind of information, its storage and its replay.
6. Justify the statement that nucleic acids are used for both information transfer and information storage.
7. State similarities and differences between the nervous and hormone systems. Compare the form of information transfer with cellular communication or with communication between individuals.
8. Why is it incorrect to refer to the sounds of "talking" parrots as a language?
9. Discuss the function of signalling colours during the courtship of many animals. Compare this with the function of warning colours.

● Think about connections   ● See connections   ● Develop connections

Basic Concepts 117

Glossary

Accommodation [əˌkɒməˈdeɪʃn]
The adjustment of the eye to an object at varying distances e.g. by changing the refraction power of the eye lens.

Action, conditioned [kənˌdɪʃnd ˈækʃn]
An association between a behavioural element and an activated drive that has been acquired by learning.

Action potential (AP) [ˌækʃn pəˈtenʃl] [ˌeɪˈpiː]
A brief but measurable change in the membrane potential when an axon is activated. It propagates along the axon by changing ion channel conductivity.

Adaptation [ˌædəpˈteɪʃn]
The adjustment, especially of sensory organs or receptors, to different intensities of a stimulus, e.g. in the eye by changing/adjusting the iris.

Afferent [ˈæfərənt]
In nerves: conveying (impulses) from the sensory cells towards the central nervous system.

Aggression [əˈgreʃn]
A collective term for attacking, defending and threatening behaviour within a species or between species.

All-or-nothing principle [ˌɔːlɔːˈnʌθɪŋ ˌprɪnsɪpl]
A principle applying to certain processes that either occur completely and with maximal intensity or not at all, e.g. the *action potential* in nerve cells.

Altruism [ˈæltrʊɪzm]
Selfless behaviour for the benefit of others; opposite of egoism (forms: reciprocal and nepotistic altruism).

Antagonist [ænˈtægənɪst]
Opposing force, e.g. in muscles, hormones, receptors at synapses and chemical reactions.

Axon [ˈæksɒn]
A long extension (up to 100cm) of a nerve cell that transmits impulses from the cell body to the synapses.

Bateman principle [ˈbeɪtmən ˌprɪnsɪpl]
The reproductive success of males increases with every copulation partner; this is not true for females.

Behavioural ecology [bɪˌheɪvjərəl ɪˈkɒlədʒɪ]
Science of the adaptation of behavioural patterns (not only of social behaviour) to the ecological framework.

Behaviourism [bɪˈheɪvjərɪzm]
An approach in learning psychology (originated in America) that reduces behaviour to simple stimulus-reaction schemata.

Causation, proximate [ˌprɒksɪmət kɔːˈzeɪʃn]
Genetic, physiological (e.g. hormone) or any other direct influencing factor that causes the development of a feature.

Causation, ultimate [ˌʌltɪmət kɔːˈzeɪʃn]
Reason that determines the development of a feature from the view of selection theory. Hereby, the function of the feature and the selection value are examined.

Chronobiology [ˌkrɒnəʊbaɪˈɒlədʒɪ]
Examines the reason and function of the periodical / temporal organization of organisms.

Cognition [kɒgˈnɪʃn]
In psychology, cognition is the ability to cope with life. It includes: perception, analysis, memory, reasoning, decision making and the creation of memories.

Conditioning [kənˈdɪʃənɪŋ]
Simple learning process in which a particular reaction (reflex or action) is produced as a response to a specific stimulus.

Conduction, saltatory [ˌsɒltətɔːrɪ kənˈdʌkʃn]
Axons without myelin-sheath carry out continuous conduction. In contrast, the action potential jumps from one node of Ranvier to the next in myelinated axons exhibiting saltatory conduction.

Conflict (biological definition) [ˈkɒnflɪkt]
Occurs in an individual if motivations of the same degree exist for two opposing behavioural patterns. A conflict between two individuals occurs if the behaviour of one partner increases its own reproductive fitness but at the same time lowers the reproductive fitness of the other partner.

Depolarisation [diːˌpəʊləraɪˈzeɪʃn]
Rapid change in the resting potential (see *resting potential*).

Displacement activity [dɪˈspleɪsmənt ækˌtɪvətɪ]
Term derived from classical ethology. It describes the occurrence of a behaviour in a "misplaced" functional combination (e.g. during cock fights, there is generally neither an attack nor flighting, but pecking movements).

Effector [ɪˈfektə]
Effector organ that carries out the reaction of an animal to stimuli (muscle or gland).

Efferent [ˈefərənt]
Conveying information via nerves away from the central nervous system to effector organs such as muscles or glands.

Ethology [ɪˈθɒlədʒɪ]
The study of behaviour; a branch of biology involving the study of animal and human behaviour.

Fertilization [ˌfɜːtɪlaɪˈzeɪʃn]
The fusion of the nuclei of two gametes resulting in a fertilized egg cell (zygote); after *insemination*.

Final action (consummatory act) [ˌfaɪnl ˈækʃn] [ˌkɒnsəˌmeɪtərɪ ˈækt]
A term mainly characterized by KONRAD LORENZ as the final part of instinctive behaviour.

Fitness, reproductive [ˌriːprədʌktɪv ˈfɪtnəs]
A way to measure the reproductive success during a lifetime of an animal, i.e. the ability to pass on its own genes to its offspring.

Ganglion cell [ˈgæŋglɪən ˌsel]
A nerve cell (see *neuron*) occurring within a group (ganglion) outside the central nervous system.

Gene [dʒiːn]
A hereditary unit consisting of a sequence of DNA that occupies a specific location on a chromosome and codes for an RNA molecule.

Generalization [ˌdʒenrəlaɪˈzeɪʃn]
Learnt connection by which previous information or experience is transferred to similar situations or stimuli combinations.

Glossary

Habitat [ˈhæbɪtæt]
Environmental area that is preferred by a particular species.

Habituation [həˌbɪtjʊˈeɪʃn]
Stimulus-specific absence of a reaction or perception e. g. if the same scents are always present.
AND: In regard to learning processes: when previously shown reactions disappear.

Hormones [ˈhɔːməʊnz]
Are substrates of the secondary information system. They are chemical messengers that are produced in specific cells or organs (endocrine glands) and are transported throughout the body to their target cells by the blood.

Hyperpolarisation [ˌhaɪpəˌpəʊləraɪˈzeɪʃn]
A change in the cell's membrane potential that makes it more negative. This change is the result of increased differences in charges between the intra- and extracellular space.

Imprinting [ˈɪmprɪntɪŋ]
Learning process restricted to a sensitive developmental phase. It results in a highly stable learnt behavioural pattern.

Infanticide [ɪnˈfæntɪsaɪd]
Killing of the offspring of one's own species (mainly by males).

Inhibition, conditioned [kənˌdɪʃnd ˌɪnhɪˈbɪʃn]
Inhibition of a behaviour attributable to a negative experience (see *learning*).

Inhibition, lateral [ˌlætərəl ˌɪnhɪˈbɪʃn]
Occurs when an excited neuron reduces the activity of its neighbours within a neuronal network. This helps to increase contrasts, e. g. the edges between light and dark areas.

Injurious fight [ɪnˌdʒʊərɪəs ˈfaɪt]
High-risk fighting behaviour in which the animals use harmful weapons such as teeth, claws, and the pointed ends of antlers. A deadly ending is possible (see *ritualized fight*).

Instinctive behaviour [ɪnˌstɪŋktɪv bɪˈheɪvjə]
Term introduced by KONRAD LORENZ and keenly discussed today. It describes a largely constant and "congenital" behavioural pattern that is triggered by a response mechanism to key stimuli. It only occurs if there is a specific readiness to act.

Key stimulus [ˌkiː ˈstɪmjələs]
Term in classical ethology describing a stimulus or stimulus combination that triggers a genetically determined action (see *instinctive behaviour*).

Learning [ˈlɜːnɪŋ]
Change in behaviour or character as a result of individual experience. Learning is based on changes within neuronal structures.

'Lock and key' model [ˌlɒkənˈkiː ˌmɒdl]
Two molecules possess specific complementary geometric shapes that fit exactly into one another, e. g. enzyme and substrate.

Membrane potential [ˌmembreɪn pəˈtenʃl]
Electrical voltage caused by the different distributions of ions across the cell membrane. In an axon, it alternates between the *resting* and *action potential*.

Memory [ˈmemərɪ]
Storage within the brain of experienced / acquired information. This information can be recalled as memories.

Monogamy [məˈnɒgəmɪ]
If only one female and one male form a couple.

Nerve [nɜːv]
A bundle of nerve fibres surrounded by connective tissue. It transmits electrical impulses (see *action potential*).

Neuron (nerve cell) [ˈnjʊərɒn]
A specialized cell type that generates and transmits electrical impulses.

Parasympathetic nervous system [ˌpærəsɪmpəˌθetɪk ˈnɜːvəsˌsɪstəm]
Part of the autonomic (vegetative) nervous system that reduces the activity of, for example, muscles and stimulates digestion; antagonist to the sympathetic nervous system.

Parental investment [pəˌrentl ɪnˈvestmənt]
Parental care that is invested into raising offspring. It increases the probability of survival of the offspring but it partly prevents the parents from producing more offspring.

Phytohormone [ˌfaɪtəʊˈhɔːməʊn]
Plant messenger substance that controls plant development (e. g. germination, growth, seed maturation, losing leaves). Plants, in contrast to animals, have no real morphologically distinct hormone glands.

Polyandry [ˈpɒlɪændrɪ]
If one female lives together with several males (see *mating system*).

Polygamy [pəˈlɪgəmɪ]
If one individual lives together with several individuals of the opposite sex.

Polygynandry [ˌpɒlɪdʒɪnˈændrɪ]
If several females live together with several males (see *mating system*).

Polygyny [pəˈlɪdʒɪnɪ]
If one male lives with several females (see *mating system*).

Population [ˌpɒpjəˈleɪʃn]
Group of individuals of one species living within a defined area and forming a reproductive community.

Potential, postsynaptic [ˌpəʊstsɪˌnæptɪk pəˈtenʃl]
The potential across the membrane on the postsynaptic side of the synaptic cleft can be changed to an EPSP (excitatory postsynaptic potential) or an IPSP (inhibitory postsynaptic potential) by transmitters. EPSPs are depolarizations and support the development of action potentials at the axon hillock. IPSPs are hyperpolarizations and inhibit the generation of action potentials.

Receptive field [rɪˌseptɪv ˈfiːld]
All receptors of a sensory organ that are primarily connected to one sensory neuron.

Receptor [rɪˈseptə]
Sensory physiology: a zone in which impulses are triggered by specific impulses; molecular biology: most often a membrane-bound molecule that binds specific molecules and, as a result, triggers processes within a cell.

Reflex [ˈriːfleks]
To a large extent, a genetically determined, involuntary reaction of an animal to a particular stimulus.

Glossary

Reflex arc [ˈriːfleks ˌɑːk]
Chain of processes and structures (receptors, afferent and efferent nerve tracts, synapses, reflex centres and effectors) that are responsible for the course of a reflex.

Refractory period [rɪˈfræktərɪ ˌpɪərɪəd]
Period of time after an impulse in which an excitable membrane (e.g. of an axon) cannot generate new impulses, despite a depolarization, because certain ion channels are inactive.

Regulation [ˌreɡjəˈleɪʃn]
Maintenance of a certain state despite changes in influencing factors. The compensation of errors and the generation of equilibrium are mainly performed via negative feedback loops.

Repolarisation [riːˌpəʊlərarˈzeɪʃn]
The return to the resting potential (membrane potential) after an action potential has passed by (opposite of *depolarisation*).

Resting potential [ˌrestɪŋ pəˈtenʃl]
Electrical voltage difference between the inside and the outside of an excitable membrane in a non-excited state, e.g. on nerve or muscle cells.

Ritualized fight [ˌrɪtjʊəlaɪzd ˈfaɪt]
Low-risk fighting behaviour between animals of the same species. The animals either do not use injurious weapons or use them only in a ritualized manner without injuring the opponent, e.g. by threatening, pushing.

Sensitive phase [ˈsensətɪv ˌfeɪz]
A restricted period of time in which specific behavioural patterns are established by obligatory learning process (e.g. *imprinting*).

Sensory cell [ˈsensərɪ ˌsel]
Cells in which specific stimuli trigger neuronal impulses.

Signal [ˈsɪɡnl]
Stimulus that enables the transmission of encoded information.

Social system [ˈsəʊʃl ˌsɪstəm]
Describes the way that males and females live together (compare *mating system*).

Sociobiology [ˌsəʊʃɪəʊbarˈɒlədʒɪ]
Science that studies the biological adaptation of social behaviour of animal and human populations.

Sodium-potassium pump (Na^+/K^+ pump)
[ˌsəʊdɪəmpəˌtæsɪəm ˈpʌmp]
ATP-consuming transport mechanism that lies in the cell membrane and that is able to transport Na^+ and K^+ ions, even against high concentration gradients. These pumps are essential for the *resting potential* of muscle and nerve cells.

Stimulus [ˈstɪmjələs]
Environmental factor that influences the organism and that can trigger impulses at specific cells (receptors). Stimulus reception and processing are fundamental characteristics of life.

Strategy [ˈstrætədʒɪ]
Genetically determined decision-making that controls the use of different behavioural alternatives of the same function in different situations.

Stress [stres]
Sum of all reactions of an organism to increased environmental demands.

Sympathetic nervous system
[ˌsɪmpəˌθetɪk ˈnɜːvəs ˌsɪstəm]
Part of the autonomic (vegetative) nervous system that activates certain organs (e.g. skeletal muscles, heart) to increase the body's effectiveness and that inhibits others (e.g. digestion, sexual functions). Antagonist to the *parasympathetic nervous system*.

Synapse [ˈsaɪnæps]
Junction of impulse transmission between a nerve cell on the one side and muscle, nerve or gland cells on the other side.

Synaptic cleft [sɪˌnæptɪk ˈkleft]
Space between the presynaptic membrane of the synaptic knob of one cell and the postsynaptic membrane of the receiving cell.

Synaptic knob (bouton)
[sɪˌnæptɪk ˈnɒb] [buːˈtɒn]
A swelling at the end of an axon.

Tactic [ˈtæktɪk]
Behavioural alternatives of same function within a strategy.

Taxis [ˈtæksɪs]
Orienting movement by freely moving organism; the direction of movement is decided by the direction of the stimulus. If the movement is directed towards the stimulus, it is called positive taxis; if the movement occurs in the opposite direction, it is called negative taxis.

Territory [ˈterɪtrɪ]
An area that is defended by an organism against members of the same species in order to secure a resource.

Tradition [trəˈdɪʃn]
Transfer of learnt information within a group and between generations.

Index

A

abscisic acid 60 [æbˌsɪzɪk 'æsɪd]
abstraction 87 [æb'strækʃn]
accidental signalling 108 [ˌæksɪdentl 'sɪgnəlɪŋ]
accommodation 28 [əˌkɒmə'deɪʃn]
acetylcholine 18, 19, 20, 39 [əˌsi:tl'kəʊli:n]
acetylcholinesterase 20 [əˌsi:tlˌkəʊlɪ'nestəreɪz]
actin 19 ['æktɪn]
action potential 12, 14, 16 [ækʃn pə'tenʃl]
action-specific energy 69 [ˌækʃnspəˌsɪfɪk 'enədʒɪ]
actual value 32, 55 [ˌæktʃʊəl 'vælju:]
acultative learning 82 [əˌkʌltətɪv 'lɜ:nɪŋ]
adaptation 32 [ˌædəp'teɪʃn]
addiction 47 [ə'dɪkʃn]
adequate stimulus 26 [ˌædɪkwət 'stɪmjələs]
adrenalin 39, 55 [ə'drenəlɪn]
adrenocorticotropic hormone (ACTH) 59 [əˌdri:nəʊˌkɔ.tɪkəʊ'trɒpɪk ˌhɔ:məʊn] [ˌeɪsi:ti:'eɪtʃ]
afferent 6 ['æfərənt]
agar 60 ['eɪga:]
aggregation 93 [ˌægrɪ'geɪʃn]
aggression 106 [ə'greʃn]
alkyl phosphate 20 ['ælkɪl 'fɒsfeɪt]
all-or-nothing principle 12 [ˌɔ:lɔ:'nʌθɪŋ ˌprɪnsɪpl]
alpine chough 87 ['ælpaɪn 'tʃʌf]
altruism 101 ['æltrʊɪzm]
amacrine 31 ['æməkrɪn]
amacrine cell 29 ['æməkrɪn ˌsel]
Ammophila pubescens 77 [əˌmɒfɪlə pju:'beskenz]
amplitude code 19 ['æmplɪtju:d ˌkəʊd]
amygdala 44 [ə'mɪgdələ]
antagonist 113 [æn'tægənɪst]
anterior chamber 28 [ænˌtɪərɪə 'tʃeɪmbə]
anterior horn 38 [ænˌtɪərɪə 'hɔ:n]
Aplysia 72, 78 [ə'pli:ʒə]
appetence 69 ['æpɪtəns]
Atlantic Puffin 91 [ətˌlæntɪk 'pʌfɪn]
Atropa belladonna 21 [əˌtrəʊpə ˌbelə'dɒnə]
atropine 21 ['ætrəʊpi:n]
autoimmune disorder 21 [ˌɔ:təʊɪ'mju:n dɪˌsɔ:də]
autonomic nervous system 38 [ˌɔ:tənɒmɪk 'nɜ:vəsˌsɪstəm]
auxin 61 ['ɔ:ksɪn]
axon 8 ['æksɒn]
axon hillock 8 [ˌæksɒn 'hɪlək]

B

BAERENDS, GERARD P. 77 [ˌʒerɑ:d ˌpi: 'beərəndz]
BANTING, FREDERICK 56 [ˌfredərɪk 'bæntɪŋ]
barbiturate 47 [bɑ:'bɪtjʊrət]
basilar membran 27 ['bæsɪlə ˌmembreɪn]
BATEMAN, ANGUS JOHN 95, 107 [ˌæŋgəs ˌdʒɒn 'beɪtmən]
Bateman principle 95 ['beɪtmən ˌprɪnsɪpl]
behavioural ecology 67 [bɪˌheɪvjərəl ɪ'kɒlədʒɪ]
behaviourism 68 [bɪ'heɪvjərɪzm]
BELL, DOROTHY 90 [ˌdɒrəθɪ 'bel]
BERNARD, CLAUDE 21 [ˌklɔ:d bɜ:'nɑ:d]
BEST, CHARLES 56 [ˌtʃɑ:lz 'best]
binocular visual field 29 [bɪˌnɒkjələ ˌvɪʒʊəl 'fi:ld]
biochemical signal amplification 51 [ˌbaɪəʊkemɪkl 'sɪgnl ˌæmplɪfɪˌkeɪʃn]
bipolar cell 29, 31 [baɪ'pəʊlə ˌsel]
blackbird 89 ['blækbɜ:d]
black widow 20 [blæk 'wɪdəʊ]
blind spot 29 ['blaɪnd ˌspɒt]
blink reflex 25 ['blɪŋk ˌri:fleks]
blood glucose level 54, 56, 57 [ˌblʌd 'glu:kəʊz ˌlevl]
blood sugar level 114 [ˌblʌd 'ʃʊgə ˌlevl]
blue tit 88, 95 ['blu: ˌtɪt]
bombicol 116 ['bɒmbɪkɒl]
botulinum toxin 20 [bɒtjʊ'laɪnəm ˌtɒksɪn]
BOYSEN-JENSEN, PETER 60 [ˌpi:tə ˌbɔɪzən'dʒensən]
brain death 43 ['breɪn ˌdeθ]
brain research 42 ['breɪn ˌrɪ:sɜ:tʃ]
brainstem 41 ['breɪnstem]
BREHM, ALFRED 70 [ˌælfrɪd 'breɪm]
brood parasitism 108 ['bru:d ˌpærəsɪtɪzm]
BROWN, ROBERT 110 [ˌrɒbət 'braʊn]

C

CAJAL, SANTIAGO RAMÓN Y 9 [sæntɪˌɑ:gəʊ ræˌmɒn i: kæ'hɑ:l]
causative connection 67 [ˌkɔ:zətɪv kə'nekʃn]
cell body 8 [ˌsel 'bɒdɪ]
central nervous system (CNS) 6, 38 [ˌsentrəl 'nɜ:vəsˌsɪstəm] [ˌsi:en'es]
central pit 29 [ˌsentrəl 'pɪt]
cerebellum 41 [ˌserɪ'beləm]
cerebral blood flow 42 [ˌserəbrəl 'blʌd ˌfləʊ]
cerebral cortex 40, 44 [ˌserəbrəl 'kɔ:teks]
cerebral hemisphere 40 [ˌserəbrəl 'hemɪsfɪə]
cerebrum 40 [ˌserəbrəm]
chemoreceptor 26 ['ki:məʊrɪˌseptə]
Chinese monal 109 [ˌtʃaɪni:z 'məʊnəl]
choice of partner 95 [ˌtʃɔɪs əv 'pɑ:tnə]
choroid 28 ['kɒrɔɪd]
chronobiology 74 [ˌkrɒnəʊbaɪ'ɒlədʒɪ]
chronometry 74 [krə'nɒmətrɪ]
ciliary body 28 [ˌsɪlɪərɪ 'bɒdɪ]
ciliary muscle 28 [ˌsɪlɪərɪ 'mʌsl]
circadian rhythm 74 [sɜ:ˌkeɪdɪən 'rɪðm]
cis form 30 ['sɪs ˌfɔ:m]
classical conditioning 80 [ˌklæsɪkl kən'dɪʃənɪŋ]
cloaca 98 [kləʊ'ɑ:kə]
closed anonymous society 93 [ˌkləʊzd ə'nɒnɪməs səˌsaɪətɪ]
closed non-anonymous society 93 [ˌkləʊzd ˌnɒnə'nɒnɪməs səˌsaɪətɪ]
Clostridium botulinum 20 [klɒˌstrɪdɪəm bɒtjʊ'laɪnəm]
coal tit 88 ['kəʊl ˌtɪt]
cochlea 27 ['kɒklɪə]
cognition 37 [kɒg'nɪʃn]
common pheasant 109 [ˌkɒmən 'feznt]
common slipper shell 103 [ˌkɒmən 'slɪpə ˌʃel]
communication 108, 116 [kəˌmju:nɪ'keɪʃn]
compound eye 28 ['kɒmpaʊnd 'aɪ]
concentration gradient 11 [ˌkɒnsən'treɪʃn ˌgreɪdɪənt]
conditional strategy 96 [kənˌdɪʃnl 'strætədʒɪ]
conditioned action 80 [kənˌdɪʃənd 'ækʃn]
conditioned appetence 80 [kənˌdɪʃənd 'æpɪtəns]
conditioned inhibition 80 [kənˌdɪʃənd ˌɪnhɪ'bɪʃn]
conditioned stimulus 80 [kənˌdɪʃənd 'stɪmjələs]
conditioned strategy 96 [kənˌdɪʃənd 'strætədʒɪ]
conditioning 80 [kən'dɪʃənɪŋ]
conduction speed 17 [kən'dʌkʃn ˌspi:d]
cone 29 [kəʊn]
conflict 97 ['kɒnflɪkt]
conformational change 115 [ˌkɒnfɔ:ˌmeɪʃənl 'tʃeɪndʒ]
confusion effect 92 [kən'fju:ʒn ɪˌfekt]

coniine 20 ['kəʊniːn]
Conium maculatum 20 [ˌkəʊniəm mækjʊˈlɑːtəm]
consummatory act 69 [ˌkɒnsəˌmeɪtəriˌˈækt]
continuous conduction 14 [kənˌtɪnjʊəs kənˈdʌkʃn]
contrast 34 [ˈkɒntrɑːst]
control 114 [kənˈtrəʊl]
controlled variable 55 [kənˈtrəʊld ˌveərɪəbl]
controller 32 [kənˈtrəʊlə]
control loop 114 [kənˈtrəʊlˌluːp]
control variable 32 [kənˈtrəʊl ˌveərɪəbl]
cooperative signalling 108 [kəʊˌɒprətɪv ˈsɪgnəlɪŋ]
cornea 28 [ˈkɔːnɪə]
corpus callosum 40 [ˌkɔːpəs kəˈləʊzəm]
correcting variable 55 [kəˈrektɪŋ ˌveərɪəbl]
cortex 40 [ˈkɔːteks]
countercurrent 113 [ˈkaʊntəˌkʌrənt]
crossed extensor reflex 25 [ˌkrɒst ɪkˈstensə ˌriːfleks]
cup of hemlock 20 [ˌkʌp əv ˈhemlɒk]
cupula 27 [ˈkʌpjʊlə]
curare 21 [kjʊˈrɑːrɪ]
CUSHING, HARVEY 42 [ˌhɑːvɪ ˈkʊʃɪŋ]
cyclic adenosine monophosphate 51 [ˌsaɪklɪk əˌdenəsiːn ˌmɒnəʊˈfɒsfeɪt]
cytokinin 60 [ˌsaɪtəʊˈkaɪnɪn]

D

DARWIN, CHARLES 60, 68, 101 [ˌtʃɑːlz ˈdɑːwɪn]
deception 109 [dɪˈsepʃn]
deceptive signalling 108 [dɪˌseptɪv ˈsɪgnəlɪŋ]
decoy 70 [ˈdiːkɔɪ]
dendrite 8 [ˈdendraɪt]
depolarisation 12 [diːˌpəʊlərarˈzeɪʃn]
development 52 [dɪˈveləpmənt]
diabetes mellitus 56 [daɪəˌbiːtiːz məˈlaɪtəs]
direct fitness 101 [daɪˌrekt ˈfɪtnəs]
dishabituation 78 [ˌdɪshəbɪtjʊˈeɪʃn]
displacement activity 69 [dɪˈspleɪsmənt ækˌtɪvəti]
distress 59 [dɪˈstres]
dorsal root 38 [ˌdɔːsl ˈruːt]
drive theory of aggression 69 [ˈdraɪv ˌθɪəri əv əˈgreʃn]
Drosophila 95 [drɒˈsɒfɪlə]
dunnock 98 [ˈdʌnək]
dwarf mongoose 92 [ˌdwɔːf ˈmɒŋguːs]

E

ear 27 [ɪə]
ecstasy 47 [ˈekstəsɪ]
effector 7, 24, 32, 55 [ɪˈfektə]
efferent 6 [ˈefərənt]
electroencephalography 42 [ɪˌlektrəʊˌensefəˈlɒgrəfi]
electrogenic pump 11 [ɪˌlektrəʊˌdʒenɪk ˈpʌmp]
electroreceptor 26 [ɪˈlektrəʊrɪˌseptə]
electrotonic conduction 14 [ɪˌlektrəʊˌtɒnɪk kənˈdʌkʃn]
elephant seal 94 [ˈeləfənt ˌsiːl]
endorphin 46 [enˈdɔːfɪn]
epinephrine (US) 39, 55 [ˌepɪˈnefrɪn]
epiphysis 41 [ɪˈpɪfɪsɪs]
escalation strategy 104 [ˌeskəˈleɪʃn ˌstrætədʒɪ]
Ethology 67 [ɪˈθɒlədʒɪ]
ethylene 61 [ˈeθɪliːn]
eustress 59 [ˈjuːstres]
evolutionarily stable 96 [ˌiːvəˌluːʃənərɪli ˈsteɪbl]
excitatory postsynaptic potential (EPSP) 18, 22, 42 [ekˌsaɪˌteɪtəri ˌpəʊstsɪˌnæptɪk pəˈtenʃl] [ˌiːpiːesˈpiː]
excitatory synapse 22 [ekˌsaɪˌteɪtəri ˈsaɪnæps]

extinction 80 [ɪkˈstɪŋkʃn]
extracellular recording 10 [ˌekstrəˌseljələ rɪˈkɔːdɪŋ]
eye 28 [aɪ]

F

falling phase 12 [ˈfɔːlɪŋ ˌfeɪz]
feedback 49 [ˈfiːdbæk]
feticide 100 [ˈfiːtɪsaɪd]
fighting strategy 104 [ˈfaɪtɪŋ ˌstrætədʒɪ]
fight-or-flight syndrome 59 [ˌfaɪtɔːˈflaɪt ˌsɪndrəʊm]
final action 69 [ˌfaɪnl ˈækʃn]
fitness 94 [ˈfɪtnəs]
fitness increase 69 [ˌfɪtnəs ˈɪnkriːs]
fixed action pattern 68, 69 [ˌfɪkst ˈækʃn ˌpætən]
floater 89 [ˈfləʊtə]
Florida Scrub Jay 101 [ˌflɒrɪdə ˈskrʌb ˌdʒeɪ]
food imprinting 84 [ˈfʊt ˌɪmprɪntɪŋ]
fovea 29 [ˈfəʊvɪə]
frequency code 19 [ˈfriːkwənsi ˌkəʊd]
FREUD, SIGMUND 36 [ˌsɪgmənd ˈfrɔɪd]
FRISCH, KARL VON 68 [ˌkɑːl vɒn ˈfrɪʃ]

G

GALVANI, LUIGI 10 [luːˌiːdʒi gælˈvɑːnɪ]
ganglion cell 29 [ˈgæŋglɪən ˌsel]
gathering-dispersal society 106 [ˌgæðərɪŋdɪˈspɜːsl səˌsaɪəti]
gender conflicts 95 [ˈdʒendə ˌkɒnflɪkts]
gene 72 [dʒiːn]
general adaptation syndrome (GAS) 59 [ˌdʒenrəl ædæpˈteɪʃn ˌsɪndrəʊm] [ˌdʒiːeɪˈes]
generalization 87 [ˌdʒenrəlaɪˈzeɪʃn]
gibberellin 60 [ˌdʒɪbəˈrɪlɪn]
Giemsa solution 9 [ˈgiːmzə səˌluːʃn]
glial cell 8 [ˈgliːəl ˌsel]
glucagon 55 [ˈgluːkəgɒn]
glucocorticoid 59 [ˌgluːkəʊˈkɔːtɪkɔɪd]
glucose value 56 [ˈgluːkəʊz ˌvæljuː]
GOLGI, CAMILLO 9 [kəˈmɪləʊ ˈgɒldʒɪ]
GOODALL, JANE 106 [ˌdʒeɪn ˈgʊdɔːl]
great tit 102 [ˌgreɪtˌtɪt]
grooming 107 [ˈgruːmɪŋ]
guarding monogamy 99 [ˌgɑːdɪŋ məˈnɒgəmi]
GUDERNATSCH, JOHN F. 48 [ˌdʒɒn ef ˈgʊdənætʃ]
guppy 102 [ˈgʌpɪ]

H

habitat 88 [ˈhæbɪtæt]
habitat choice 88 [ˈhæbɪtætˌtʃɔɪs]
habitat imprinting 84 [ˈhæbɪtæt ˌɪmprɪntɪŋ]
habituation 78 [həˌbɪtjʊˈeɪʃn]
hair follicle receptor 27 [ˈheə ˌfɒlɪkl rɪˌseptə]
Hamilton's inequality 101 [ˌhæmɪltənz ˌɪnɪˈkwɒləti]
handicap principle 109 [ˈhændɪkæp ˌprɪnsɪpl]
Hanuman Langur 100 [ˌhʌnʊmɑːn lʌŋˈgʊə]
HASSENSTEIN, BERNHARD 81 [ˌbɜːnhɑːd ˈhæsənstaɪn]
HEINROTH, OSKAR 68, 84 [ˌɒskə ˈhaɪnrɒθ]
HELLER 91 [ˈhelə]
HELMHOLTZ, HERMANN VON 33 [ˌhɜːmən vɒn ˈhelmhəʊlts]
HERING, EWALD 33 [ˌjuːwəld ˈherɪŋ]
hermit crab 102 [ˈhɜːmɪt ˌkræb]
Herring Gull 103 [ˈherɪŋ ˌgʌl]
HESS, ECKHARD H. 84 [ˌekhɑːd ˌeɪtʃ ˈhes]
hierarchy 94, 106 [ˈhaɪərɑːki]
hippocampus 41 [ˌhɪpəʊˈkæmpəs]

HODGKIN, ALAN 12	[ˌælən ˈhɒdʒkɪn]	**M**	
HOLST, DIETRICH VON 58	[ˌdiːtrɪk vɒn ˈhɒlst]	magnetic resonance imaging 43	[mægˌnetɪk ˌrezənəns ˈɪmɪdʒɪŋ]
horizontal cell 29, 34	[ˌhɒrɪˈzɒntl ˌsel]	mechanoreceptor 26, 27	[ˈmekənəʊrɪˌseptə]
hormone 48, 52, 117	[ˈhɔːməʊn]	median canal 27	[ˈmiːdɪən kəˌnæl]
hyperpolarisation 12, 22	[ˌhaɪpəˌpəʊləraɪˈzeɪʃn]	medulla oblongata 41	[mɪˌdʌlə ˌɒblɒŋˈgɑːtə]
hypoglycaemia 57	[ˌhaɪpəʊglaɪˈsiːmɪə]	Meissner's corpuscle 27	[ˌmaɪsnəz ˈkɔːpəsəl]
hypophysis 41, 48	[haɪˈpɒfɪsɪs]	membrane potential 10	[ˌmembreɪn pəˈtenʃl]
hypothalamus 41, 58	[ˌhaɪpəʊˈθæləməs]	MERING, JOSEPH VON 56	[ˌdʒəʊsəf vɒn ˈmɪərɪŋ]
		Merkel nerve ending 27	[ˌmɜːkl ˈnɜːv ˌendɪŋ]
I		Merkel's disc 27	[ˈmɜːklz ˈdɪsk]
imitation 86	[ˌɪmɪˈteɪʃn]	metamorphosis 48, 50	[ˌmetəˈmɔːfəsɪs]
imprinting of breeding parasites 84	[ˌɪmprɪntɪŋ əv ˌbriːdɪŋ ˈpærəsaɪts]	midbrain 41	[ˈmɪdbreɪn]
indirect fitness 101	[ˌɪndaɪˌrekt ˈfɪtnəs]	MILINSKI 91	[mɪˈlɪnskɪ]
indole-3-acetic acid (IAA) 61	[ˌɪndəʊlˌθriːəˌsiːtɪk ˈæsɪd]	MINKOWSKI, OSKAR 56	[ˌɒskə mɪŋˈkɒfskɪ]
infanticide 100	[ɪnˈfæntɪsaɪd]	monarch butterfly 66	[ˌmɒnək ˈbʌtəflaɪ]
information 116	[ˌɪnfəˈmeɪʃn]	monocular visual field 29	[məˌnɒkjələ ˌvɪʒʊəl ˈfiːld]
information flow 19	[ˌɪnfəˈmeɪʃn ˌfləʊ]	monogamy 93, 98	[məˈnɒgəmɪ]
inhibitory postsynaptic potential (IPSP) 22	[ɪnˌhɪbɪtərɪ ˌpəʊstsɪˌnæptɪk pəˈtenʃl] [ˌaɪpiːesˈpiː]	monosynaptic 24	[ˌmɒnəʊsɪˈnæptɪk]
inhibitory synapse 22	[ɪnˌhɪbɪtərɪ ˈsaɪnæps]	monosynaptic reflex 25	[ˌmɒnəʊsɪˈnæptɪk ˌriːfleks]
injurious fight 94	[ɪnˌdʒʊərɪəs ˈfaɪt]	mood 46	[muːd]
injurious fighting 104	[ɪnˌdʒʊərɪəs ˈfaɪtɪŋ]	motor 6	[ˈməʊtə]
instinct 69	[ˈɪnstɪŋkt]	motor end plate 19	[ˌməʊtər ˌend ˈpleɪt]
instinct conditioned interconnection 69	[ˌɪnstɪŋktkənˌdɪʃnd ˌɪntəkəˈnekʃn]	motor nerve 24	[ˈməʊtə ˌnɜːv]
instinctive behaviour 69	[ɪnˌstɪŋktɪv bɪˈheɪvjə]	muscle spindle 16, 24	[ˈmʌsl ˌspɪndl]
insulin 54	[ˈɪnsjʊlɪn]	*Myasthenia gravis* 21	[maɪəsˌθiːnɪə ˈgrævɪs]
interference variable 55	[ˌɪntəˈfɪərəns ˌveərɪəbl]	myelin 8	[ˈmaɪəlɪn]
interneuron 38, 78	[ˌɪntəˈnjʊərɒn]	myelinated axon 15	[ˌmaɪəlɪneɪtɪd ˈæksɒn]
ion channel 11	[ˈaɪən ˌtʃænl]	myelin sheath 8	[ˌmaɪəlɪn ˈʃiːθ]
ion theory of excitation 13	[ˌaɪən ˌθɪərɪ əv ˌeksaɪˈteɪʃn]	myosin 19	[ˈmaɪəsɪn]
iris 28	[ˈaɪrɪs]		
irreversible 84	[ˌɪrɪˈvɜːsəbl]	**N**	
islets of Langerhans 54	[ˌaɪləts əv ˈlæŋəhæns]	Na⁺/K⁺ pump 11	[ˌsəʊdɪəmpəˌtæsɪəm ˈpʌmp]
		negative feedback 32	[ˌnegətɪv ˈfiːdbæk]
J		NEHER, ERWIN 15	[ˌɜːwɪn ˈneɪhə]
Jarman-Bell principle 90, 93	[ˌdʒɑːmənˈbel ˌprɪnsɪpl]	neostigmine 21	[ˌniəʊˈstɪgmiːn]
JARMAN, PETER 90	[ˌpiːtə ˈdʒɑːmən]	nerve 8	[nɜːv]
jetlag 74	[ˈdʒetlæg]	nerve fibre 8	[ˈnɜːv ˌfaɪbə]
		nervous system 38	[ˈnɜːvəs ˌsɪstəm]
K		neurobiology 6	[ˌnjʊərəʊbaɪˈɒlədʒɪ]
Kaspar-Hauser animal 82	[ˌkæspəˈhaʊzər ˌænɪml]	neuron 8, 17	[ˈnjʊərɒn]
KATZ, BERNARD 12	[ˌbɜːnəd ˈkæts]	neuronal switch 22	[ˌnjʊərənl ˈswɪtʃ]
key stimulus 69	[ˌkiː ˈstɪmjələs]	neurosecretion 48	[ˌnjʊərəʊsɪˈkriːʃn]
knee jerk reflex 24	[ˈniːdʒɜːk ˌriːfleks]	neurotoxin 20	[ˌnjʊərəʊˈtɒksɪn]
KOEHLER, OTTO 87	[ˌɒtəʊ ˈkɜːlə]	neurotransmitter 18	[ˌnjʊərəʊtrænzˈmɪtə]
KÖHLER, WOLFGANG 86	[ˌwʊlfgæŋ ˈkɜːlə]	night blindness 30	[ˈnaɪt ˌblaɪndnəs]
		nodes of Ranvier 15	[ˌnəʊdz əv ˈrɒnvɪeɪ]
L		noradrenalin 39	[ˌnɔːrəˈdrenəlɪn]
labyrinth 27	[ˈlæbərɪnθ]	norepinephrine (US) 39	[ˌnɔːrepɪˈnefrɪn]
lateral inhibition 34	[ˌlætərəl ˌɪnhɪˈbɪʃn]		
Latrodectus 20	[ˌlætrəˈdektəs]	**O**	
law of dual quantification 69	[ˌlɔː əv ˌdjuːəl ˌkwɒntɪfɪˈkeɪʃn]	obligatory learning 82, 84	[əˌblɪgətrɪ ˈlɜːnɪŋ]
learning 82, 86, 87	[ˈlɜːnɪŋ]	odour 27	[ˈəʊdə]
LEHRMANN, DANIEL S. 63	[ˌdænjəl ˌes ˈleəmən]	olfactory sensory neuron 27	[ɒlˌfæktərɪ ˈsensərɪ ˌnjʊərɒn]
Lilian's lovebird 73	[ˌlɪlɪənz ˈlʌvbɜːd]	open anonymous society 93	[ˌəʊpn əˈnɒnɪməs səˌsaɪətɪ]
limbic system 39, 41, 44	[ˈlɪmbɪk ˌsɪstəm]	opium 46	[ˈəʊpɪəm]
location imprinting 84	[ləʊˈkeɪʃn ˌɪmprɪntɪŋ]	opsin 30	[ˈɒpsɪn]
lock-and-key model 112	[ˌlɒkənˈkiː ˌmɒdl]	optic disc 29	[ˌɒptɪk ˈdɪsk]
Loligo 12	[ləʊˈliːgəʊ]	optic nerve 29	[ˌɒptɪk ˈnɜːv]
long-term memory 44	[ˌlɒŋtɜːm ˈmemərɪ]	oral glucose tolerance test 54	[ˌɔːrəl ˌgluːkəʊz ˈtɒlərəns ˌtest]
LORENZ, KONRAD 68, 69, 82, 84	[ˌkɒnræd ˈlɔːrənz]	organ of Corti 27	[ˌɔːgən əv ˈkɔːtɪ]
lymph 11	[lɪmf]	overshoot 12	[ˌəʊvəˈʃuːt]
lysosome 50	[ˈlaɪsəsəʊm]	oxytocin 62	[ˌɒksɪˈtəʊsɪn]

Index **123**

P

PAAL, ARPAD 60 [ˌɑːpæd ˈpɑːl]
Pacinian corpuscle 27 [pəˌsɪniən ˈkɔːpəsəl]
pain fibre 46 [ˈpeɪn ˌfaɪbə]
papillae 27 [pəˈpɪliː]
PARACELSUS 46 [ˌpærəˈselsəs]
parasympathetic nervous system 38 [ˌpærəsɪmpəˌθetɪk ˈnɜːvəsˌsɪstəm]
parental investment 97 [pəˌrentl ɪnˈvestmənt]
patch-clamp technique 15 [ˈpætʃklæmp tekˌniːk]
PAWLOW, IWAN PETROWITSCH 80 [ˌaɪvən ˌpetrəvɪtʃ ˈpævlɒf]
peacock pheasant 109 [ˈfeznt ˌpiːkɒk]
perception 35 [pəˈsepʃn]
peripheral nervous system 38 [pəˌrɪfərəl ˈnɜːvəsˌsɪstəm]
pheromone 116 [ˈferəməʊn]
photoreceptor 26 [ˈfəʊtəʊrɪˌseptə]
phototropism 60 [ˌfəʊtəʊˈtrəʊpɪzm]
Phytohormone 60 [ˈfaɪtəʊˌhɔːməʊn]
pineal gland 41 [ˈpɪniəl ˌglænd]
pit eye 26 [ˈpɪt ˌaɪ]
pituitary gland 48 [pɪˈtjuːɪtəri ˌglænd]
PLATO 20 [ˈpleɪtəʊ]
polarisation 28 [ˌpəʊləraɪˈzeɪʃn]
polyandry 93, 98 [ˈpɒliændri]
polygamy 93 [pəˈlɪgəmi]
polygynandry 93, 98 [ˌpɒlɪdʒɪnˈændri]
polygyny 93, 98 [pəˈlɪdʒəni]
polysynaptic reflex 25 [ˌpɒlɪsɪˈnæptɪk ˌriːfleks]
pons 41 [pɒnz]
positron emission tomography 43 [ˌpɒsɪtrɒn ɪˌmɪʃn təˈmɒgrəfi]
posterior horn 38 [pɒsˌtɪəriə ˈhɔːn]
postsynaptic 18 [ˌpəʊstsɪˈnæptɪk]
potassium (K) 11 [pəˈtæsɪəm]
preset 115 [ˌpriːˈset]
presynaptic 18 [ˌpriːsɪˈnæptɪk]
presynaptic inhibition 22 [ˌpriːsɪˌnæptɪk ˌɪnhɪˈbɪʃn]
primary colour 33 [ˌpraɪməri ˈkʌlə]
primary sensory cell 16 [ˌpraɪməri ˈsensəri ˌsel]
principle 113 [ˈprɪnsɪpl]
protease 50 [ˈprəʊtɪeɪs]
proximate causation 64, 67 [ˌprɒksɪmət kɔːˈzeɪʃn]
pupil 28 [ˈpjuːpəl]

R

receiver 108 [rɪˈsiːvə]
receptive field 34 [rɪˌseptɪvˌˈfiːld]
receptor 26, 32 [rɪˈseptə]
receptor cell 16 [rɪˈseptəˌsel]
receptor potential 16 [rɪˌseptə pəˈtenʃl]
reconciliation 107 [ˌrekənˌsɪliˈeɪʃn]
reflex 24, 78 [ˈriːfleks]
reflex arc 24 [ˈriːfleksˌɑːk]
refraction power 28 [rɪˈfrækʃnˌpaʊə]
refractory period 13, 14 [rɪˈfræktəriˌpɪəriəd]
regulation 55, 114 [ˌregjəˈleɪʃn]
releasing hormone 41 [rɪˈliːsɪŋˌhɔːməʊn]
repolarisation 12 [riːˌpəʊləraɪˈzeɪʃn]
reproductive fitness 100 [ˌriːprəˌdʌktɪvˌˈfɪtnəs]
reproductive strategies 102 [ˌriːprəˈdʌktɪvˌstrætədʒɪz]
residual reproductive value 103 [rɪˌzɪdjʊəlˌriːprəˈdʌktɪvˌvælju:]
resting potential 10, 11, 17 [ˌrestɪŋ pəˈtenʃl]

Rete mirabile 113 [ˌriːtɪ mɪˈrɑːbɪleɪ]
retina 29, 30, 33 [ˈretɪnə]
retinal pigment epithelium 28 [ˌretɪnəlˌpɪgmənt epɪˈθiːlɪəm]
Rhabditis inermis 72 [ræbˌdaɪtɪs ɪˈnɜːmɪs]
rhodopsin 30 [rəʊˈdɒpsɪn]
rising phase 12 [ˈraɪzɪŋˌfeɪz]
ritualisation 109 [ˌrɪtjʊəlaɪˈzeɪʃn]
ritualized fighting 104 [ˌrɪtjʊəlaɪzd ˈfaɪtɪŋ]
rod 29 [rɒd]
rosy-faced lovebird 73 [ˌrəʊzɪfeɪst ˈlʌvbɜːd]

S

SAKMANN, BERT 15 [ˌbɜːt ˈsækmən]
saltatory conduction 15 [ˌsɒltətəri kənˈdʌkʃn]
sarin 20 [ˈsærɪn]
satellite male 96 [ˈsætəlaɪtˌmeɪl]
SCHONS 98 [ˈʃɒnz]
sclera 28 [ˈsklɪərə]
second messenger 31, 51 [ˌsekənd ˈmesɪndʒə]
sensitive phase 84 [ˈsensətɪvˌfeɪz]
sensitization 83 [ˌsensɪtaɪˈzeɪʃn]
sensor 55 [ˈsensə]
sensor for rotation 27 [ˌsensə fə rəʊˈteɪʃn]
sensory 6 [ˈsensəri]
sensory cell 28 [ˈsensəriˌsel]
sensory memory 44 [ˌsensəri ˈmeməri]
sensory nerve 24, 26 [ˈsensəriˌnɜːv]
sensory organ 26 [ˈsensəriˌɔːgən]
sensory receptor 28 [ˌsensəri rɪˈseptə]
set value 32, 55, 114 [ˌset ˈvæljuː]
short-term memory 44 [ˌʃɔːtːɜːmˌˈmeməri]
signal 108 [ˈsɪgnl]
simulation 105 [ˌsɪmjʊˈleɪʃn]
sinoatrial node 25 [ˌsaɪnəʊˈeɪtriəlˌnəʊd]
sinus node 25 [ˈsaɪnəsˌnəʊd]
SKINNER, BURRHUS F. 80 [ˌbʌrəsˌef ˈskɪnə]
sleep research 43 [ˈsliːpˌrɪsɜːtʃ]
social hierarchy 106 [ˈsəʊʃl ˈhaɪərɑːki]
social system 93 [ˈsəʊʃlˌsɪstəm]
sociobiology 65, 67 [ˌsəʊʃɪəʊbaɪˈɒlədʒi]
SOCRATES 20 [ˈsɒkrətiːz]
sodium (Na) 11 [ˈsəʊdɪəm]
sodium-potassium pump 11 [ˌsəʊdɪəmpəˌtæsɪəm ˈpʌmp]
Solanaceae 21 [ˌsəʊləˈneɪsiiː]
soma 8 [ˈsəʊmə]
SPALDING, DOUGLAS 84 [ˌdʌgləsˌˈspɔːldɪŋ]
spatial summation 23 [ˌspeɪʃl sʌmˈeɪʃn]
spinal cord 24, 38 [ˈspaɪnlˌkɔːd]
spinal ganglia 38 [ˌspaɪnl ˈgæŋgliə]
stimulating hormone 41 [ˈstɪmjəleɪtɪŋˌhɔːməʊn]
stimulus processing 33 [ˈstɪmjələsˌprəʊsesɪŋ]
stimulus, stimuli 26, 33, 77 [ˈstɪmjələs, ˈstɪmjəliː]
strategy 96, 98, 102, 104 [ˈstrætədʒi]
stress 58 [stres]
stressor 59 [ˈstresə]
stress situation 58 [ˈstresˌsɪtjʊˌeɪʃn]
surface extension 113 [ˌsɜːfɪs ɪkˈstenʃn]
Suxamethonium 20 [ˌsʌksəmeˈθəʊnɪəm]
sympathetic nervous system 38 [ˌsɪmpəˌθetɪk ˈnɜːvəsˌsɪstəm]
synapse 8, 18, 22 [ˈsaɪnæps]
synaptic cleft 18 [sɪˌnæptɪkˌˈkleft]
synaptic knob 8, 18 [sɪˌnæptɪkˌˈnɒb]
synaptic vesicle 18 [sɪˌnæptɪk ˈvesɪkl]

T
tabun 20 [təˈbuːn]
tactic 96 [ˈtæktɪk]
taste bud 27 [ˈteɪst ˌbʌd]
taxis 69 [ˈtæksɪs]
temporal summation 23 [ˌtempərəl sʌmˈeɪʃn]
territory 89 [ˈterɪtrɪ]
testicular feminization 53 [tesˌtɪkjələ ˌfemɪnaɪˈzeɪʃn]
thalamus 41 [ˈθæləməs]
thaumatin 56 [ˈθɔːmətɪn]
Thaumatococcus daniellii 56 [ˌθɔːmətəˌkɒkəs dænˈjelɪiː]
theory of instinctive behaviour 68 [ˌθɪərɪ əv ɪnˌstɪŋktɪv bɪˈheɪvjə]
thermoreceptor 26 [ˈθɜːməʊrɪˌseptə]
THORNDIKE, EDWARD L. 68, 80 [ˌedwəd ˌel ˈθɔːndaɪk]
thyroid 48 [ˈθaɪrɔɪd]
thyroxin 48 [θaɪˈrɒksiːn]
thyroxin-releasing factor (TRF) 48 [θaɪˌrɒksiːnrɪˈliːsɪŋ ˌfæktə ˌtiːɑːrˈef]
thyroxin-stimulating hormone (TSH) 48 [θaɪˌrɒksiːnˈstɪmjəleɪtɪŋ ˌhɔːməʊn] [ˌtiːesˈeɪtʃ]
TINBERGEN, NIKOLAAS 68 [ˌnɪkələs ˈtɪnbɜːgən]
tomography 43 [təˈmɒgrəfɪ]
tool usage 86 [ˈtuːl ˌjuːzɪdʒ]
tradition 86, 110 [trəˈdɪʃn]
transduction 26 [trænzˈdʌkʃn]
trans form 30 [trænsˈfɔːm]
transmitter 108 [trænzˈmɪtə]
trichromatic colour theory 33 [ˌtraɪkrəʊˌmætɪk ˈkʌlə ˌθɪərɪ]
TRIVERS, ROBERT L. 97 [ˌrɒbət ˌel ˈtraɪvəz]
truthfulness 108 [ˈtruːθfəlnəs]

tubocurarine 21 [ˌtjuːbəʊkʊˈrɑːriːn]
tupaia 58 [tʊˈpaɪə]
tympanic canal 27 [tɪmˈpænɪk kəˌnæl]

U
ultimate causation 65, 67 [ˌʌltɪmət kɔːˈzeɪʃn]
unconditioned response 80 [ˌʌnkəndɪʃənd rɪˈspɒns]
universalism 111 [ˌjuːnɪˈvɜːsəlɪzm]

V
vacuum activity 69 [ˈvækjuːm ækˌtɪvətɪ]
vegetative nervous system 38 [ˌvedʒɪtətɪv ˈnɜːvəs ˌsɪstəm]
ventral root 38 [ˌventrl ˈruːt]
ventricle 42 [ˈventrɪkl]
visual field 29, 36 [ˌvɪʒʊəl ˈfiːld]
vitreous body 28 [ˌvɪtrɪəs ˈbɒdɪ]

W
WAAL, FRANS DE 107 [ˌfræns də ˈwɑːl]
WATSON, JOHN B. 68 [ˌdʒɒn ˌbiː ˈwɒtsən]
WENT, FRITS 61 [ˌfrɪts ˈwent]
WILLIS, THOMAS 56 [ˌtɒməs ˈwɪlɪs]
WILTSCHKO, WOLFGANG 75 [ˌwʊlfgæŋ ˈwɪltʃkəʊ]

Y
YOUNG, THOMAS 33 [ˌtɒməs ˈjʌŋ]

Z
ZIPPELIUS, HANNA MARIA 85 [ˌhænə məˌrɪə zɪˈpelɪəs]
zonular fibre 28 [ˈzəʊnjələ ˌfaɪbə]

Pictures sources

U1 Jupiterimages GmbH (IFA/Krämer), Ottobrunn/München; **6.1** FOCUS (Science Photo Library), Hamburg; **7.1** Okapia (Howie Garber), Frankfurt; **7.2** Corbis (zefa/E. & P. Bauer), Düsseldorf; **8.2** Cedric S. Raine, Bronx; **9.1; 9.2** Johannes Lieder, Ludwigsburg; **9.3** Prof. Dr. Manfred Keil, Neckargemünd; **9.4** Cajal Legacy. Instituto Cajal. Madrid; **9.S** Okapia (D. Kunkel), Frankfurt; **10.1** Okapia (Norbert Lange), Frankfurt; **15.K** MPI- Medizinische Forschung (Bert Sakmann, Ernst Neher), Heidelberg; **17.S** Okapia (D. Kunkel), Frankfurt; **18.2** Okapia (Fawcett, Friend/Science Source), Frankfurt; **19.K** FOCUS (Science Pictures Limited/SPL), Hamburg; **20.1a** Okapia (John Mitchell/OSF), Frankfurt; **20.1b** FOCUS (Dr. Gary Gaugler/SPL), Hamburg; **20.1c** Okapia (Ernst Schacke, Naturbild), Frankfurt; **21.S** Okapia (Hans Reinhard), Frankfurt; **22.Rd.** Kage Mikrofotografie, Lauterstein; **26.1** Okapia (Manfred Uselmann), Frankfurt; **27.S** Prof. Jürgen Wirth, Dreieich; **29.1** Johannes Lieder, Ludwigsburg; **30.1** Okapia (NAS/Omikron), Frankfurt; **31.1a; 31.1b** Okapia (NAS/Omikron), Frankfurt; **35.S** creativ collection Verlag GmbH, Freiburg; **42.1** FOCUS (Alexander Tsiaras, Science Photo Library), Hamburg; **42.2** Picture Press (Volker Hinz/STERN), Hamburg; **42.S** Okapia (D. Kunkel), Frankfurt; **43.1** Volker Steger (M. Raichle), München; **45.1; 45.4; 45.9; 45.12; 45.13** Prof. Jürgen Wirth, Dreieich; **45.2** MEV Verlag GmbH, Augsburg; **45.3** Reinhard-Tierfoto, Heiligkreuzsteinach; **45.5** Corbis (Digital Art), Düsseldorf; **45.6** Avenue Images GmbH RF (Thinkstock), Hamburg; **45.7; 45.10** Corbis (Tom & Dee Ann McCarthy), Düsseldorf; **45.8** Avenue Images GmbH RF (Brand X Pictures), Hamburg; **45.11** Photodisc; **46.1** Mauritius Images (Zak), Mittenwald; **46.2** National Institute on Drug Abuse (Michael J. Kuhar), Baltimore, USA; **46.S** Mauritius Images (K. Paysan), Mittenwald; **54.1a; 54.1b** Okapia (Jeffrey Telner), Frankfurt; **56.1** Bilderberg (Milan Horacek), Hamburg; **56.2** Mauritius Images (Hubatka), Mittenwald; **56.S** Okapia (Jeffrey Telner), Frankfurt; **57.1** VISUM Foto GmbH (Laureen Greenfield), Hamburg; **60.1** Visuals Unlimited (Sylvan Wittwer), Hollis; **60.2** Visuals Unlimited (Jack Bostrack), Hollis; **62.1** Okapia (Tom Vezo), Frankfurt; **62.2** Lothar Lenz Pferdefotoarchiv, Dohr; **63.S** Okapia (Frank Krahmer), Frankfurt; **64.1** Corbis (Reuters), Düsseldorf; **64.2** Okapia (Prof. Bernhard Grzimek), Frankfurt; **64.3** Corbis (Sygma), Düsseldorf; **64.4** FOCUS (SPL), Hamburg; **65.1** Getty Images (Taxi/Benjamin Shearn), München; **65.2** Okapia (David Thompson/OSF), Frankfurt; **65.3** Tierbildarchiv Angermayer (Günter Ziesler), Holzkirchen; **65.4** Okapia (Manfred Danegger), Frankfurt; **65.5** Getty Images (Image Bank/Joseph Van Os), München; **65.6** Okapia (Jef Meul/SAVE), Frankfurt; **66.1** Prof. Jürgen Wirth, Dreieich; **66.Rd.1** WILDLIFE Bildagentur GmbH (J. Cox), Hamburg; **66.Rd.2** Okapia (Dr. Gilbert S. Grant), Frankfurt; **68.1** Picture-Alliance (dpa), Frankfurt; **68.2** Okapia (Darek Karp/Naturbild), Frankfurt; **69.S** Okapia (Regis Cavignaux, BIOS), Frankfurt; **71.S** Reinhard-Tierfoto, Heiligkreuzsteinach; **74.1** TOPICMedia (Lehmann), Ottobrunn; **75.S** Corel Corporation Deutschland, Unterschleissheim; **77.1; 77.S** TOPICMedia (Günter Roland), Ottobrunn; **78.1** Naturfotografie Frank Hecker (Frieder Sauer), Panten-Hammer; **79.S** TOPICMedia (Heinrich König), Ottobrunn; **80.Rd.** Mauritius Images (Science Source), Mittenwald; **81.S** Harald Lange Naturbild, Bad Lausick; **82.Rd.** ARDEA London Limited (Brian Bevan), London; **85.1; 85.2; 85.3** Hanna-Maria Zippelius, Mechernich; **85.S** Okapia (Barrie E. Watts/OSF), Frankfurt; **86.1** Prof. Dr. Jürgen Lethmate, Ibbenbüren; **86.2** Photoshot (NHPA/ORION PRESS), Hamburg; **86.Rd.** WILDLIFE Bildagentur GmbH (P. Ryan), Hamburg; **87.1; 87.6** Prof. Dr. Jürgen Lethmate, Ibbenbüren; **87.S** Okapia (NAS, Tim Davis), Frankfurt; **88.1; 88.2** Tierbildarchiv Angermayer (Rudolf Schmidt), Holzkirchen; **90.1a** TOPICMedia (J. & C. Sohns), Ottobrunn; **90.1b** Okapia (Daryl & Shama Balfour), Frankfurt; **90.1c** TOPICMedia (Sohns), Ottobrunn; **90.1d** Okapia (Frank Krahmer), Frankfurt; **90.1e** TOPICMedia (J. & C. Sohns), Ottobrunn; **91.2** WILDLIFE Bildagentur GmbH (Delpho), Hamburg; **92.1** Okapia (Joe McDonald), Frankfurt; **92.Rd.** Okapia (Sohns), Frankfurt; **93.1** Okapia (Konrad Wothe), Frankfurt; **94.1** TOPICMedia (E. u. D. Hosking), Ottobrunn; **95.1** Tierbildarchiv Angermayer (Rudolf Schmidt), Holzkirchen; **96.Rd.** Manfred Pforr Naturbild-Archiv, Langenpreising; **97.K** Gesellschaft für Primatologie e.V. (Andreas Ploß), Göttingen; **98.1** TOPICMedia (Roger Wilmshurst), Ottobrunn; **98.S** Okapia (P. Laub), Frankfurt; **100.1** Prof. Volker Sommer, Holzhausen/Rhw.; **100.Rd.** Mauritius Images (age), Mittenwald; **101.1** Tierbildarchiv Angermayer (Günter Ziesler), Holzkirchen; **102.1** TOPICMedia (Kelvin Aitken), Ottobrunn; **102.2** Okapia (Robert Maier), Frankfurt; **102.5** Okapia (Jef Meul/SAVE), Frankfurt; **102.S** Tierbildarchiv Angermayer (Sigi Köster), Holzkirchen; **103.2** Quedens, Georg, Norddorf-Amrum; **103.4** Okapia (NAS/A. Martinez), Frankfurt; **106.1** Okapia (NAS, Tom McHugh), Frankfurt; **107.K** Tierbildarchiv Angermayer (Günter Ziesler), Holzkirchen; **107.Rd.** Arco Images GmbH (P. Wegner), Lünen; **108.1a** Reinhard-Tierfoto, Heiligkreuzsteinach; **108.1b** TOPICMedia (Volkmar Brockhaus), Ottobrunn; **108.1c** Okapia (Ludwig Werle), Frankfurt; **108.2** TOPICMedia (Gerard Lacz), Ottobrunn; **110.1** Corbis (Peter Guttmann), Düsseldorf; **110.2** EPD (Anja Kessler), Frankfurt; **110.3a; 110.3b; 110.4** Okapia (NAS/Art Wolfe), Frankfurt; **110.5** EPD, Frankfurt; **110.S** Corel Corporation Deutschland, Unterschleissheim; **111.1** EPD (Höria), Frankfurt; **111.2** Prof. Jürgen Wirth, Dreieich; **111.3** Corbis (Peter Harholdt), Düsseldorf; **111.4; 111.5** Klett-Archiv (Hans-Peter Krull), Stuttgart; **111.6** Okapia (Fritz Pölking), Frankfurt; **111.7** Okapia (G. Wiltsie, Peter Arnold), Frankfurt; **111.8** Corbis (Joel Sartore), Düsseldorf; **111.9** Okapia (M. Schneider/UNEP/Still Pictures), Frankfurt; **111.10** Corbis (David Muench), Düsseldorf; **112.1** FOCUS (Andrew Syred, SPL), Hamburg; **112.2** Okapia (Fawcett, Friend/ScienceSource), Frankfurt; **112.3** FOCUS (SPL/CNRI), Hamburg; **113.1** Okapia (Manfred P. Kage), Frankfurt; **114.1; 114.2** Okapia (Ulla Spiegel), Frankfurt; **114.3** Tilman Wischuf Tier- und Naturfotografie, Stabbestad; **116.1; 116.2** Tierbildarchiv Angermayer (Hans Pfletschinger), Holzkirchen; **116.3** FOCUS (Science Photo Library/SPL), Hamburg; **117.1** Corbis (Paul Barton), Düsseldorf

Nicht in allen Fällen war es uns möglich, den Rechteinhaber der Abbildungen ausfindig zu machen. Berechtigte Ansprüche werden selbstverständlich im Rahmen der üblichen Vereinbarungen abgegolten.